Finite Element Analysis for Biomedical Engineering Applications

Finite Element Analysis for Biomedical Engineering Applications

Z. Yang

CRC Press
Taylor & Francis Group
Boca Raton London New York

CRC Press is an imprint of the
Taylor & Francis Group, an **informa** business

CRC Press
Taylor & Francis Group
6000 Broken Sound Parkway NW, Suite 300
Boca Raton, FL 33487-2742

First issued in paperback 2023

© 2019 by Taylor & Francis Group, LLC
CRC Press is an imprint of Taylor & Francis Group, an Informa business

No claim to original U.S. Government works

ISBN 13: 978-1-03-265391-4 (pbk)
ISBN 13: 978-0-367-18218-2 (hbk)
ISBN 13: 978-0-429-06126-4 (ebk)

DOI: 10.1201/9780429061264

**Visit the Taylor & Francis Web site at
http://www.taylorandfrancis.com**

**and the CRC Press Web site at
http://www.crcpress.com**

Contents

PART II   Soft Tissues

Preface

In 2001, I came to the University of Pittsburgh to pursue my PhD. As I learned about biomechanics, I became fascinated by the complications of biology. In the past 17 years, I had been working on many bioengineering projects with professors from the University of Pittsburgh, University of Pennsylvania, Allegheny General Hospital, and Soochow University. My long-time research has given me experience in finite element modeling in the field of biomedical studies. I have chosen to record my experiences in a book which, I hope, will encourage medical researchers to do further investigations. Yet, even after 17 years of study and research, I recognize that I still have more to learn about biomechanics. Should this book, therefore, contain errors, I ask readers to point them out to me so that I can address and correct them.

While I wrote this book, I received help and encouragement from many of my friends, including Frank Marx, Dr. J.S. Lin, Dr. Richard Debski, and Fayan Xu. Dr. Zhi-Hong Mao reviewed the whole manuscript. I am grateful for his constructive comments that have greatly improved the quality of the book. I give a special thanks to Ronna Edelstein for her time and effort in revising my manuscript. I express my great appreciation to the staff at CRC Press, especially Marc Gutierrez and Kari Budyk for their assistance in publishing the book. Finally, I thank my family, especially my wife, Peng, and my two children, for their constant support.

About the Author

Z. Yang earned a PhD in mechanical engineering at the University of Pittsburgh in 2004. Over the last 17 years, he has collaborated with professors from various colleges, such as the University of Pennsylvania and University of Pittsburgh, and finished a number of biomedical projects. Currently, he is a senior software engineer in the field of finite element analysis with over 10 years' experience.

1 Introduction

Because people are living longer in today's world, more individuals are dealing with a variety of diseases. Some common diseases are associated with the mechanical states of human organs. For example, hips often break when older people fall, and the lumbar disc degenerates due to excessive loadings over the long term. An abdominal aortic aneurysm (AAA) occurs when the stresses of the AAA wall exceed the strength of the wall tissue. Treatment of these diseases requires an understanding of the stress-states of relevant parts under various conditions. When some parts of the human body degenerate and lose their function, people may have to undergo implant surgeries, such as stent implantation for treatment of atherosclerosis and total knee replacement to regain the walking function. Although these implants can improve the person's quality of life significantly, they can also raise other issues, such as medial tilting in ankle replacements and fatigue and wear of the liner in hip implants. To solve these issues and improve the medical designs, it is vital to study the mechanical behavior of the implants.

While researchers are testing the mechanical responses of the organs and the implants in the lab, they also emphasize numerical simulations, especially finite element analysis. Since the 1970s, some well-known commercial finite element codes, such as ANSYS, NASTRAN, MARC, ABAQUS, LSDYNA, and COMSOL, have been developed to solve the structural problems. Among them, ANSYS software has the most powerful nonlinear solver, and hence it has become the most widely used software in both academia and industry. Over the past decade, many advanced finite element technologies have been developed in ANSYS. The purpose of this book is to simulate some common medical problems using finite element advanced technologies, which paves a path for medical researchers to perform further studies.

The book consists of four main parts. Each part begins by presenting the structure and function of the biology, and then it introduces the corresponding ANSYS advanced features. The final discussion highlights some specific biomedical problems simulated by ANSYS advanced features.

The topic of Part I is bone. After this introductory chapter, Chapter 2 introduces the structure and material properties of bone. Chapter 3 discusses the nonhomogeneous character of bone, including modeling it by computed tomography (CT) in Section 3.1 and by multidimensional interpolation in Section 3.2. Chapter 4 describes how to build a finite element model of anisotropic bone, and the crack-growth in the microstructure of cortical bone is simulated by eXtended Finite Element Model (XFEM) in Chapter 5.

Part II, which deals with soft tissues, is very detailed. Chapter 6 introduces the structure and material properties of soft tissues like cartilage, ligament, and intervertebral discs (IVDs). Next, Chapter 7 presents the nonlinear behavior of soft tissues and simulation of AAA in ANSYS190. Chapter 8 examines the viscoelasticity of soft tissues, including its application to the study of periodontal ligament creep.

Some soft tissues are enhanced by fibers. Chapter 9 discusses three approaches of fiber enhancement in ANSYS190: (1) standard mesh-dependent fiber enhancement, in which the fibers are created within the regular base mesh; (2) mesh-independent fiber enhancement that creates fibers independent of the base mesh; and (3) the anisotropic material model with fiber enhancement. The first two approaches are utilized to simulate the fibers in the annulus of the intervertebral disc (IVD).

Many nonlinear material models in ANSYS are available for the simulation of soft tissues. If the experimental data of one biological material do not fit any of these models, the researchers may turn to USERMAT in ANSYS. Chapter 10 focuses on the topic of how to develop user material models in ANSYS.

The soft tissues are biphasic, consisting of 30%–70% water. Chapter 11 introduces ways of modeling soft tissues as porous media and the application of biphasic modeling in head impact and IVD creep research.

Part III describes joint simulation. After briefly introducing the structure of joints in Chapter 12, in the next chapter, Section 13.1 defines three contact types in a whole-knee simulation, and a two-dimensional (2D) axisymmetrical poroelastic knee model is built in Section 13.2. Then, the discrete element method of knee joint that is implemented in ANSYS is analyzed in Chapter 14.

Part IV presents a number of implant simulations. Chapter 15 studies the contact of the talar component and the bone to investigate medial tilting in ankle replacement. The stent implantation is simulated in Chapter 16 using the shape memory alloy super-elasticity model. The Archard wear model is applied to study the wear of the hip implant in Chapter 17. Chapter 18 predicts the fatigue life of a mini-dental implant using ANSYS SMART technology.

Chapter 19 presents a retrospective look at the entire content of the book. Some guidelines are summarized for the simulation of biomedical problems.

The biomedical problems in this book have been simulated using ANSYS Parametric Design Language (APDL). Reading this book requires knowledge of APDL. To learn APDL, I suggest first reading the ANSYS help documentation and then practice some technical demonstration problems available in this documentation. All APDL input files of the finite element models in the book are provided in the appendixes.

Part I

Bone

An adult human body has 206 separate bones, which generate red and white blood cells, reserve minerals, support the body, and allow mobility. Clinical study of bone indicates that some bone diseases such as osteoporosis and bone surgeries like total hip replacement surgery are associated with the bone, which requires an understanding of the mechanical stresses in human bones. Therefore, Part I focuses on the finite element modeling of bone.

Chapter 2 introduced the structure and material properties of bone. Next, Chapter 3 presented two approaches to study nonhomogeneous bone. The anisotropic bone model was built in Chapter 4, and crack growth in the cortical bone was studied in Chapter 5 using the eXtended Finite Element Method (XFEM).

2 Bone Structure and Material Properties

The material properties of bone are closely associated with bone structure. Therefore, in this chapter, bone structure is introduced first, followed by a description of the material properties of bone.

2.1 BONE STRUCTURE

Bone consists of both fluid and solid elements. Water comprises up to 25% of the total weight of bone. Roughly 85% of this water is scattered in the organic matrix around the collagen fibers and ground substances, and the other 15% is found in canals and cavities within the bone cells. The solid part of bone is composed of organic and inorganic components. The organic material, known as *collagen,* composes about 95% of the extracellular matrix and accounts for about one-third of bone's dry weight. The inorganic material, hydroxyapatite, or $Ca_{10}(PO4)_6(OH)_2$, consists of the minerals calcium and phosphate and accounts for the rest of the dry weight of bone.

The inorganic materials—apatite crystals—are very stiff and strong, with the Young's modulus of 165 GPa, compared with the Young's modulus of steel (200 GPa), and aluminum (6061 alloy, 70 GPa). Apatite crystals are embedded into the collagen to form a composite material. Such a bone structure has a Young's modulus between that of apatite and collagen. It also has a very high bone strength because, like concrete, collagen gives bone its flexibility, while the inorganic materials impart its resilience.

Figure 2.1 illustrates the cancellous (spongy) and cortical (compact) bone. The thin outer layer is the periosteum. The cortical bone completely covers the periosteum. From there, the cortical bone transits radically to the cancellous bone. The central part is the marrow. The cortical bone is approximately four times the mass of the cancellous bone. Both bones have the same basic materials, but the differences between them are the degree of porosity and the organization. The porosity of the cortical bone ranges from 5%–30%, while the cancellous bone has a porosity from 30%–90%. The bone porosity is not constant, but it always changes based on the loadings, diseases, and the aging process.

Because bone is a living organ, it can change; for example, it can grow relevant to its stress state. Bone is assumed to have adapted to different living conditions and to have achieved the maximum-minimum design. The human femur has a remarkable adaptation of its structure to the mechanical requirements caused by the load on the femur head (Figure 2.2) [1].

5

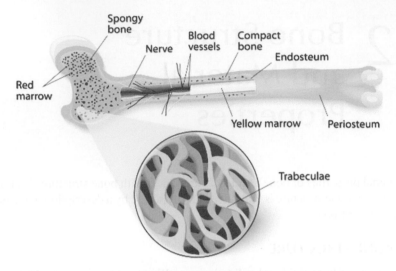

FIGURE 2.1   Structure of a femur. (Designua © 123RF.com.)

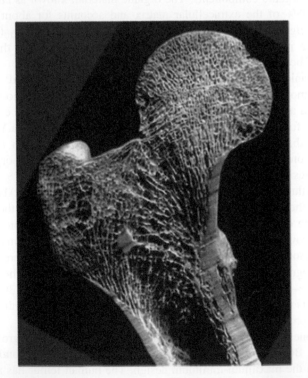

FIGURE 2.2   Photograph of the upper femur in coronal section. (From Gray, H., *Anatomy of the Human Body*. Lea & Febiger, Philadelphia, 1918 [1].)

2.2 MATERIAL PROPERTIES OF BONE

Bone has a linear relation between its stress and its strain in a certain range. The slope of this range is defined as the Young's modulus. Figure 2.3 [2] illustrates the material properties of bone compared with other materials.

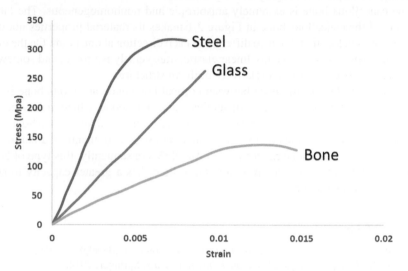

FIGURE 2.3 Comparative properties of various materials. (From Pal, S., *Design of Artificial Human Joints & Organs*. Springer, 2014 [2].)

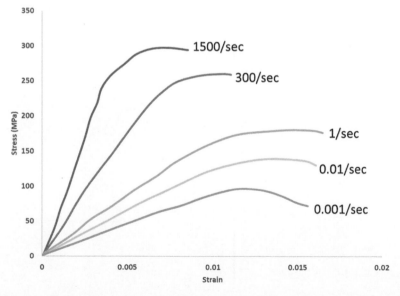

FIGURE 2.4 Strain-rate sensitivity of the cortical bone. (From Pal, S., *Design of Artificial Human Joints & Organs*. Springer, 2014 [2].)

The mineral content of the bone affects its material properties [3]. Higher mineralization increases the bone's stiffness but decreases its toughness; it reduces the bone's capability of absorbing shock and strain energy.

Bone is more strain rate sensitive than other biological tissues (Figure 2.4) [2], which may cause bone-ligament and bone-tendon injuries. The optimal strain rate for energy absorption is roughly 0.1–1 per second.

The cancellous bone is extremely anisotropic and nonhomogeneous. The lattice structure of the cancellous bone in Figure 2.1 makes its material properties not only different in each location, but also different in each direction at one point. On the other hand, the cortical bone is roughly linear elastic, transversely isotropic, and somewhat homogeneous because it has a relatively uniform structure.

Another significant difference between cortical bone and cancellous bone is that cortical bone is more than 40 times stiffer than cancellous bone, which makes the cortical part sustain greater stress, but less strain before failure. Unlike cortical bone, cancellous bone sustains strains of 75%, but much less stress. As mentioned before, the porosity of cancellous bone ranges from 30%–90%; consequently, this type of bone, which is filled with blood, marrow, and body fluid, has a greater capacity to store energy than cortical bone.

REFERENCES

1. Gray, H., *Anatomy of the Human Body*. Lea & Febiger, Philadelphia, 1918.
2. Pal, S., *Design of Artificial Human Joints & Organs*. Springer, 2014.
3. Fung, Y. C., *Biomechanics: Mechanical Properties of Living Tissues*. 2nd ed., Springer, New York, 1993.

3 Simulation of Nonhomogeneous Bone

As discussed in Section 2.1 of Chapter 2, bone is nonhomogeneous. Its material properties vary across the entire bone. Two approaches have been developed to study nonhomogeneous bone. One is building a bone model from computed tomography (CT) data. The other is to interpolate the bone material properties using multidimensional interpolation. These approaches are presented in Sections 3.1 and 3.2 of this chapter, respectively.

3.1 BUILDING BONE MODEL FROM CT DATA

The determination of the mechanical stresses in human bone is very important in both research and clinical practice because the understanding of the mechanical stresses in human bones benefits the design of prostheses and the evaluation of fracture risk. For example, after total hip replacement surgery, stresses in some regions of the remaining bone diminish because the implant carries a portion of the load, which is known as *stress shielding*. According to Wolff's law, the shielded bone remodels as a response to the changed mechanical environment, resulting in loss of bone mass through the resorption and consequent loosening of the prosthesis. To alleviate this problem, the stress distribution of the bone with the prosthesis should match that of the healthy bone as much as possible.

Finite element analysis has been widely used to study the mechanical response of the human bone because the stresses in bone cannot be measured noninvasively in vivo. With the development and advancement of CT, finite element models have been built based on bone data derived using that technology. CT data provide the bone geometry because of the high contrast between the bone's tissues and the soft tissues around the bone. This information also determines the material properties of bone, as CT numbers are nearly linearly correlated with the apparent density of biologic tissue, and empirical relations between density and mechanical properties of bone have been established [1–4]. The following list presents a general path to build a finite element model of bone:

1. Build bone geometry from CT data, and mesh it using the meshing tool.
2. Calculate the density of the bone from CT data and convert the bone density data into material properties of the bone.
3. Assign the material properties of the bone to all elements of the bone.

Next, we built one finite element model of the human femur in ANSYS190.

3.1.1   CT Data

A femur of a 45-year-old female was scanned, and CT data were obtained in standard DICOM formats. The slice thickness is 0.6 mm.

3.1.2   Finite Element Model

A stack of CT image files was imported into MIMICS to generate a three-dimensional (3D) geometry of the femur (Figure 3.1). After being transferred to ICEM, the 3D geometry was meshed into two-dimensional (2D) facet elements, and then further into 3D solid elements (Figure 3.2) comprised of 756,152 elements and 139,159 nodes. Afterward, the finite element model was saved in a file with the ANSYS format and then moved from ICEM to ANSYS190. CT data were also imported in ANSYS190 by the *VREAD command in a form of 2D array—namely, VHU (,).

3.1.3   Calculation of the Average CT Number (*HU*)

After the input of CT *HU* data into ANSYS190, an average *HU* value of each 3D solid element was calculated by [1]

$$\overline{HU}_n = \frac{\int_{V_n} HU(x, y, z)\, dV}{\int_{V_n} dV}, \tag{3.1}$$

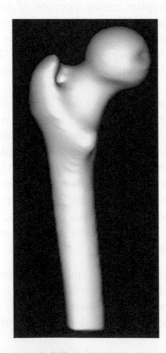

FIGURE 3.1   3D geometry created in MIMICS from a CT scan.

FIGURE 3.2   3D finite element model built in ICEM from 3D geometry. It comprises 756,152 elements and 139,159 nodes.

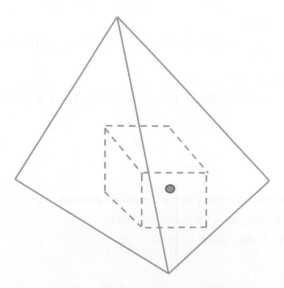

FIGURE 3.3   Schematic to compute the gray value of one element.

where V_n refers to the volume of the element n, and (x, y, z) are the coordinates in the CT reference system.

The elements are tetrahedral, with random shapes. It is impossible to determine the points within the element and complete the volume integral calculation using Eq. (3.1). Thus, an alternative was proposed—finding the average HU value in a cube with the same volume of the element and centering in the middle of the element (Figure 3.3). The center and volume of the element can be acquired by the *GET command. The APDL commands are given as follows:

```
*GET, NUM_E, ELEM, 0, COUNT        ! get number of elements

*GET, E_MIN, ELEM, 0, NUM, MIN     ! get min element number

*DO, I, 1, NUM_E, 1                ! output to ascii by looping over
                                     elements

CURR_E = E_MIN

*GET, VM, ELEM, CURR_E, VOLU       ! get the volume of the element

ES = (VM)**(1/3)/2                 ! half a size of equivalent cube

NPOINT = 0

TOTALHU = 0

*DO, J, 1, TOTPOINTS, 1

! CENTRX, CENTRY, CENTRZ is the coordinates of element center

! VHU(j,1), VHU(j,2), VHU(j,3) are the x, y, z coordinates of
point J

*IF, VHU(J,1), GT, CENTRX(CURR_E)-ES, AND, VHU(J,1), LT, CENTRX
(CURR_E)+ES, THEN

*IF, VHU(J,2), GT, CENTRY(CURR_E)-ES, AND, VHU(J,2), LT, CENTRY
(CURR_E)+ES, THEN

*IF, VHU(J,3), GT, CENTRZ(CURR_E)-ES, AND, VHU(J,3), LT, CENTRZ
(CURR_E)+ES, THEN

NPOINT = NPOINT + 1

TOTALHU = TOTALHU + VHU(J,4)       ! VHU(j,4) is HU value at point J

*ENDIF

*ENDIF

*ENDIF

*ENDDO

! array HU element saves HU value in elements

HU_ELEMENT(CURR_E) = TOTALHU/NPOINT

*GET, E_MIN, ELEM, CURR_E, NXTH

*ENDDO
```

HU values in the whole femur are plotted in Figure 3.4.

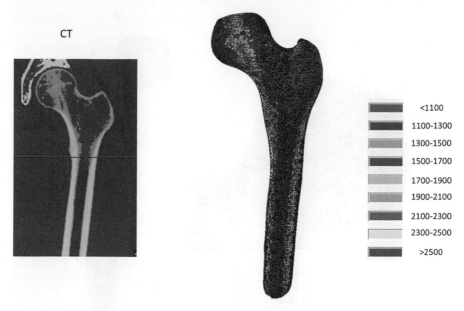

CT

■	<1100
■	1100-1300
■	1300-1500
■	1500-1700
▨	1700-1900
▨	1900-2100
■	2100-2300
▨	2300-2500
■	>2500

FIGURE 3.4   Gray values of the femur model.

3.1.4 MATERIAL PROPERTY ASSIGNMENT

The material properties of the bone, converted from the *HU* data about the bone, are based on two assumptions:

1. The relationship between CT numbers and the apparent density is linear [2]:

$$\rho_n = \alpha + \beta HU_n, \qquad (3.2)$$

where ρ_n is the average density assigned to the element n, HU_n is the average CT number, and α and β are the coefficients.

2. An empirical relationship exists between Young's modulus and the apparent bone density [3,4]:

$$E_n = a + b\rho_n^c, \qquad (3.3)$$

where E_n is the Young's modulus of element n, and a and b are the coefficients.

Theoretically, each element may have its own material identity (ID). If so, it will generate tons of material IDs in the model, which is unnecessary. For simplification, only six material properties were defined in this study for the different parts of the bone. The elements were assigned the material IDs associated with the closest Young's modulus. Figure 3.5 illustrates the material distribution.

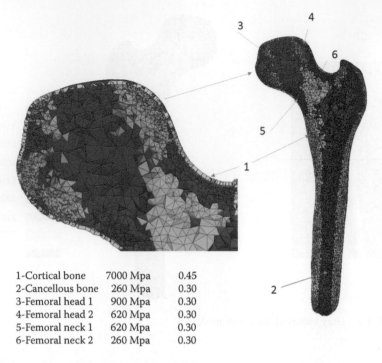

1-Cortical bone	7000 Mpa	0.45
2-Cancellous bone	260 Mpa	0.30
3-Femoral head 1	900 Mpa	0.30
4-Femoral head 2	620 Mpa	0.30
5-Femoral neck 1	620 Mpa	0.30
6-Femoral neck 2	260 Mpa	0.30

FIGURE 3.5   Material properties of the femur model.

3.1.5   DISCUSSION

The finite element model of the bone was developed in ANSYS190 from its CT data. In the determination of the CT number for each element, a simple, practical way was proposed in order to compute the CT number for each element from the equivalent cube. *HU* values in the whole femur matching CT in Figure 3.4 confirm that this new method works.

However, this method has two drawbacks. One is that the noises exist in CT data, which make the material properties neither accurate nor continuous. It is very common to see some elements with extremely high Young's moduli distributed sparsely around the bone model. Another drawback is that the equations to convert CT data to material properties of bone are purely empirical. They are correct statistically, but when applied to specific elements of the bone model, their results tend to have errors. To overcome these limitations, an interpolation of bone material properties is proposed and described in Section 3.2. In addition, the materials in the model were assumed to be isotropic. Obviously, the femur should be modeled as anisotropic to match the structure of bone. This point is addressed in Chapter 4.

3.1.6   SUMMARY

The bone model was built in ANSYS190, in which the elements were assigned different material IDs based on the CT data to represent the nonhomogeneity of the bone.

3.2 INTERPOLATION OF BONE MATERIAL PROPERTIES

Section 3.1 focused on how to build the bone model from CT data in ANSYS190. The built bone model exhibits a variation of material properties across the bone. However, it has two drawbacks, as previously discussed. To overcome these drawbacks, another approach has been explored. The material properties of the bone can be interpolated from the experimental data when data crossing the bone are available. This method involves a multidimensional interpolation. Fortunately, ANSYS190 provides such a tool, which includes the Linear Multivariate (LMUL), the Radial Basis (RBAS), and the Nearest Neighbor (NNEI) algorithms. Although multidimensional interpolation has been studied intensively in certain fields, such as medical imaging and geological modeling [5], only a few studies have been done in the area of bone research. Multidimensional interpolation, a new and advanced feature of ANSYS, makes it possible to interpolate part of the cancellous bone in the ankle. The following discussion introduces multidimensional interpolation.

3.2.1 MULTIDIMENSIONAL INTERPOLATION

As previously mentioned, three algorithms are available in ANSYS190 to perform interpolation when two or more field variables are given [6]:

- RBAS algorithm
- NNEI algorithm
- LMUL algorithm

The algorithms are described in the next subsections.

3.2.1.1 RBAS Algorithm

The RBAS algorithm uses a global algorithm to ensure C1 continuity. It is very useful for high-dimensional scatter data.

The RBAS algorithm is used to solve the following radial-basis function $Z = f(X_1, X_2, X_3 \ldots X_O)$ at all input data points:

$$Z = \sum_{i=1}^{N} a_i \left[\left(\sum_{j=1}^{O} (X_j - x_{ji})^2 \right) + c^2 \right], \qquad (3.4)$$

where N is the number of data points and O is the number of free variables.

After solving the equation, the unknown values a_i and c are obtained. Therefore, the values of query points are determined by Eq. (3.4). Eq. (3.4) is solved with all the input data provided, which require intensive computation. The advantage of this method is that the values of the query points are continuous in all areas.

3.2.1.2 NNEI Algorithm

The NNEI algorithm is very straightforward. The value of a query point equals that of the supporting point closest to the query point, which is expressed as

$$f(q) = f(x_0), \qquad (3.5)$$

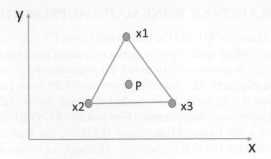

FIGURE 3.6 Schematic of the LMUL algorithm. Value of point P is interpolated by points x_1, x_2, and x_3.

where q is the query point and x_0 is the closest supporting point to the query point:

$$\|x_0 - q\| = \min (\| x - q \|) \quad x \in X, \tag{3.6}$$

where X is the set of all supporting points.

The NNEI algorithm is the fastest of these three algorithms. However, it gives relatively less accurate interpolation results.

3.2.1.3 LMUL Algorithm

Unlike the RBAS global algorithm, the LMUL algorithm uses neighbor points near the query location to perform the interpolation process. Thus, it is a local algorithm. Its basic idea is to find the neighbor points (shown as x_1, x_2, and x_3 in Figure 3.6) around the query point P, and then conduct the interpolation to obtain the value of the query point P by

$$f(\mathbf{x}) = a_1 f(\mathbf{x}_1) + a_2 f(\mathbf{x}_2) + \cdots + a_{n+1} f(\mathbf{x}_{n+1}), \tag{3.7}$$

where a_i is the shape function and $f(\mathbf{x_i})$ is the value of the neighbor point.

The LMUL algorithm requires the query point to be inside the supporting points. If the query point is outside the bounding box (Figure 3.7), its value is equal to that of the projecting point at the boundary of the bounding box. This is one of the significant drawbacks of this algorithm because in some cases, the value being outside the bounding box gives unrealistic results.

The LMUL algorithm is very fast for a large selection of data points. It becomes highly accurate when sufficient supporting points are provided. However, the query points outside the bounding box may be problematic in practice.

3.2.2 Interpolation of Material Properties of the Ankle

Jensen et al. [7] measured Young's moduli of cancellous bones throughout the ankle, including three layers at various heights and in each layer at 12 points. Therefore, Young's moduli of 36 points all over the cancellous bones (Figure 3.8) are available for interpolation. Figure 3.9 shows parts of the cancellous bones in the ankle that are

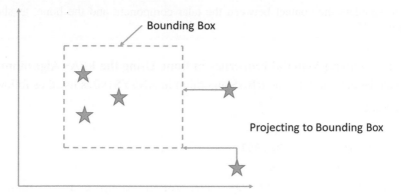

FIGURE 3.7 Schematic of the LMUL algorithm when query points are outside the bounding box.

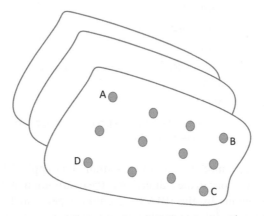

FIGURE 3.8 Schematic of Jensen's experiment, which has three layers, with 12 points in each layer. ABCD is the bounding box.

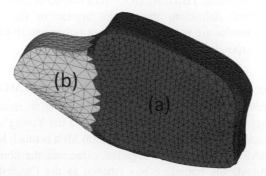

FIGURE 3.9 A part of the cancellous bone in the ankle: (a) Cancellous bone interpolated from experimental data; (b) cancellous bone with a Young's modulus of 280 MPa.

used to simulate the contact between the talar component and the bone, as shown in Chapter 15.

3.2.2.1   Defining Material Properties of Bone Using the RBAS Algorithm

The material definition by the RBAS algorithm in ANSYS190 is listed as follows:

```
TB, ELAS, 5

TBFIELD, XCOR,      191.75

TBFIELD, YCOR,      -22.951

TBFIELD, ZCOR,      135

! Young's modulus 111 MPa, Poisson's ratio 0.3 at location
(191.75, -22.951, 135)

TBDATA, 1,          111 , 0.30

......

TBFIELD, XCOR,      140.79

TBFIELD, YCOR,      -35.642

TBFIELD, ZCOR,      145

! Young's modulus 222 MPa, Poisson's ratio 0.3 at location
(140.79, -35.642, 145)

TBDATA, 1, 222 , 0.30

TBIN, ALGO, RBAS        ! RBAS interpolation of the above experi-
                          mental data
```

To plot the Young's modulus, a uniform strain 1 is applied in the z-direction. Therefore, the stress contour in the z-direction is the distribution of Young's modulus.

The Young's modulus by the RBAS algorithm appears in Figure 3.10, which clearly shows that the modulus is continuous and varies across the whole bone.

3.2.2.2   Defining Material Properties of Bone Using the NNEI Algorithm

Similarly, when the command TBIN, ALGO, NNEI is used to replace TBIN, ALGO, RBAS in the material definition, the results appear in the NNEI algorithm (Figure 3.11); these results are not continuous, but rather are divided into 36 small parts. Each part corresponds to one piece of experimental data.

3.2.2.3   Defining Material Properties of Bone Using the LMUL Algorithm

The LMUL algorithm has significantly different interpolation results. Some results (Figure 3.12) are negative, which is unrealistic because the Young's modulus cannot be negative. Moreover, the maximum value 1,898.06 MPa is much higher than that in the NNEI and RBAS algorithms. The negative value and the abnormal maximum value occurring outside the bounding box (shown as the Quadrilateral ABCD in Figure 3.8) indicate that the LMUL algorithm may generate unrealistic interpolation results outside the bounding box.

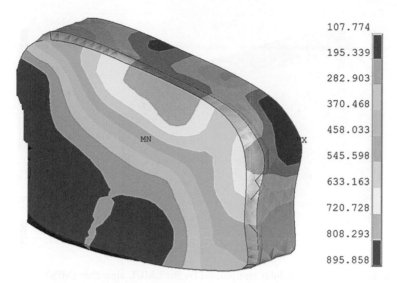

107.774
195.339
282.903
370.468
458.033
545.598
633.163
720.728
808.293
895.858

FIGURE 3.10   Young's modulus interpolated by the RBAS algorithm (MPa).

3.2.2.4   Defining Material Properties of Bone Using a Mixed Method

To avoid the abovementioned limitation of the LMUL algorithm, a mixed method was developed to combine the LMUL and NNEI algorithms, in which the LMUL algorithm is used to interpolate points inside the bounding box, and the NNEI algorithm is used for those outside the bounding box. Figure 3.13 illustrates how to sort out the points within the bounding box ABCD.

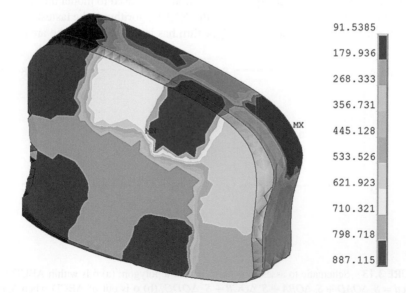

91.5385
179.936
268.333
356.731
445.128
533.526
621.923
710.321
798.718
887.115

FIGURE 3.11   Young's modulus interpolated by the NNEI algorithm (MPa).

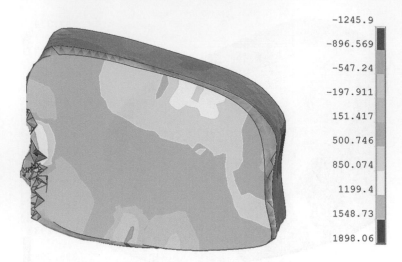

FIGURE 3.12　Young's modulus interpolated by the LMUL algorithm (MPa).

The interpolation results by this mixed method are presented in Figure 3.14; the image illustrates that the center by the LMUL algorithm is continuous and that the area outside the bounding box created by the NNEI algorithm is divided into small parts. Obviously, the fact that there are no negative and no abnormal maximum values indicates that the mixed method overcomes the limitation of the LMUL algorithm.

3.2.3　Discussion

The three multidimensional interpolation methods were used to model the cancellous bone of the ankle. Of these three methods, the NNEI algorithm is the fastest, and yet it is also the least accurate. The LMUL algorithm has issues related to the area outside

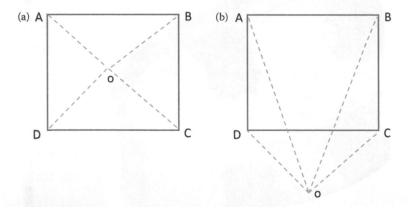

FIGURE 3.13　Schematic to assess one point within a polygon: (a) o is within ABCD when $S\_abcd = S\_\Delta OAD + S\_\Delta OBA + S\_\Delta OCB + S\_\Delta ODC$; (b) o is out of ABCD when $S\_abcd < S\_\Delta OAD + S\_\Delta OBA + S\_\Delta OCB + S\_\Delta ODC$.

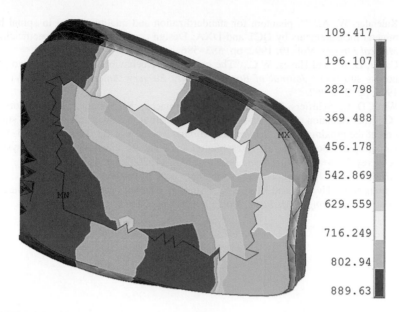

109.417

196.107

282.798

369.488

456.178

542.869

629.559

716.249

802.94

889.63

FIGURE 3.14   Young's modulus interpolated by the mixed LMUL algorithm (MPa).

the bounding box. Only the RBAS algorithm has continuous results all over the bone because it is a global interpolation, although it needs intensive computation when there are too many experimental points. Generally, the RBAS algorithm should be selected for interpolation unless doing so requires too much computational time.

The application of multidimensional interpolation methods in bone study has a theoretical base. According to Wolff's law, the bone undergoes constant changes in response to mechanical loading of the bone. Each part of the bone has a strong correlation with its adjacent parts. While it is reasonable to perform interpolation all over the bone, the material properties of the bone in Section 3.1 are determined only by CT data. The obtained material properties of one part have no connection with the other parts near it.

3.2.4   SUMMARY

Three multidimensional interpolation methods (RBAS, NNEI, and LMUL) were developed in ANSYS190. They were applied to interpolate the bone material properties. Overall, the RBAS algorithm has the best performance. Although the LMUL algorithm has an issue outside the bounding box, that problem has been resolved by a mixed method. The NNEI method is a backup for these methods.

REFERENCES

1. Taddei, F., Pancanti, A., and Viceconti, M., "An improved method for the automatic mapping of computed tomography numbers onto finite element models." *Medical Engineering & Physics*, Vol. 26, 2004, pp. 61–69.

2. Kalender, W. A., "A phantom for standardization and quality control in spinal bone mineral measurements by QCT and DXA: Design considerations and specifications." *Medical Physics*, Vol. 19, 1992, pp. 583–586.

3. Carter, D. R., and Hayes, W.C., "The compressive behaviour of bone as a two-phase porous structure." *Journal of Bone and Joint Surgery, American Volume*, Vol. 59, 1977, pp. 954–962.

4. Wirtz, D. C., Schiffers, D., Pandorf, T., Radermacher, K., Weichert, D., and Forst, R., "Critical evaluation of known bone material properties to realize anisotropic FE simulation of the proximal femur." *Journal of Biomechanics*, Vol. 33, 2000, pp. 1325–1330.

5. Amidror, I., "Scattered data interpolation methods for electronic imaging systems: A survey." *Journal of Electronic Imaging*, Vol. 11, 2002, pp. 157–176.

6. ANSYS19.0 Help Documentation in the help page of product ANSYS190.

7. Jensen, N. C., Hvid, I., and Krøner, K., "Strength pattern of cancellous bone at the ankle joint." *Engineering in Medicine*, Vol. 17, 1988, pp. 71–76.

4 Simulation of Anisotropic Bone

As mentioned in Section 2.2 of Chapter 2, bone is anisotropic. Its material properties vary not only in each location, but also in each direction at one point. This chapter introduces anisotropic material models in Section 4.1, and then presents modeling of anisotropic bone in Section 4.2.

4.1 ANISOTROPIC MATERIAL MODELS

Generally, anisotropic elastic material is specified by the following **D** matrix:

$$[D] = \begin{bmatrix} D_{11} & & & & & \\ D_{21} & D_{22} & & & & \\ D_{31} & D_{32} & D_{33} & & & \\ D_{41} & D_{42} & D_{43} & D_{44} & & \\ D_{51} & D_{52} & D_{53} & D_{54} & D_{55} & \\ D_{61} & D_{62} & D_{63} & D_{64} & D_{65} & D_{66} \end{bmatrix}. \tag{4.1}$$

The stress-strain relation can be written as either

$$\sigma = C\varepsilon \tag{4.2}$$

or

$$\varepsilon = S\sigma. \tag{4.3}$$

C in Eq. (4.2) and **S** in Eq. (4.3) are the stiffness form and compliance form of Matrix **D**, respectively.

Three commands are used to input the stiffness matrix **C** and compliance matrix **S** in ANSYS: TB, ANEL; TB, ELAS,,,, AELS (stiffness form); TB, ELAS,,,, AELF (compliance form).

The data input format is given as follows:

```
TB, ANEL or TB, ELAS,,,,AELS or TB, ELAS, ,,, AELF
TBDATA, 1, C1, C2, C3, C4, C5, C6        ! terms D11, D21, D31,
                                           D41, D51, D61
TBDATA, 7, C7, C8, C9, C10, C11, C12     ! terms D22, D32, D42,
                                           D52, D62, D33
TBDATA, 13, C13, C14, C15, C16, C17, C18 ! terms D43, D53, G63,
                                           D44, D54, D64
TBDATA, 19, C19, C20, C21                ! terms D55, D65, D66
```

The linear orthotropic material model, one special form of anisotropic material model, is defined in ANSYS using MP commands:

```
MP, EX, ,E1              ! elastic modulus, element x direction
MP, EY, ,E2              ! elastic modulus, element y direction
MP, EZ, ,E3              ! elastic modulus, element z direction
MP, PRXY/NUXY, , ν₁₂     ! major (minor) Poisson's ratio, x-y plane
MP, PRYZ/NUYZ , , ν₂₃    ! major (minor) Poisson's ratio, y-z plane
MP, PRXZ/NUXZ , , ν₁₃    ! major (minor) Poisson's ratio, x-z plane
MP, GXY, , G₁₂           ! shear modulus, x-y plane
MP, GYZ, , G₂₃           ! shear modulus, y-z plane
MP, GXZ, , G₃₁           ! shear modulus, x-z plane
```

Its **S** matrix is written as

$$[S] = \begin{bmatrix} \dfrac{1}{E_1} & -\dfrac{\nu_{21}}{E_2} & -\dfrac{\nu_{31}}{E_3} & & & \\ -\dfrac{\nu_{12}}{E_1} & \dfrac{1}{E_2} & -\dfrac{\nu_{32}}{E_3} & & 0 & \\ -\dfrac{\nu_{13}}{E_1} & -\dfrac{\nu_{23}}{E_2} & \dfrac{1}{E_3} & & & \\ & & & \dfrac{1}{G_{12}} & 0 & 0 \\ & 0 & & 0 & \dfrac{1}{G_{23}} & 0 \\ & & & 0 & 0 & \dfrac{1}{G_{13}} \end{bmatrix}. \tag{4.4}$$

Thus, it has the following compliance form of TB, ELAS:

```
TB, ELAS, , , , AELF                              ! the flexibility form
TBDATA, 1, 1/E₁, -ν₁₂/E₁, -ν₁₃/E₁, 0, 0, 0        ! terms D11, D21, D31, D41,
                                                    D51, D61
TBDATA, 7, 1/E₂, -ν₂₃/E₂, 0, 0, 0, 1/E₃           ! terms D22, D32, D42, D52,
                                                    D62, D33
TBDATA, 13, 0, 0, 0, 1/G₁₂, 0, 0                  ! terms D43, D53, G63, D44,
                                                    D54, D64
TBDATA, 19, 1/G₂₃, 0, 1/G₁₃                       ! terms D55, D65, D66
```

This flexibility form by TB, ELAS should be equivalent to the linear orthotropic model by MP commands.

4.2 FINITE ELEMENT MODEL OF FEMUR WITH ANISOTROPIC MATERIALS

As discussed in Chapter 2, the bone has a lattice structure (see Figure 2.1). Because the mechanical properties of the bone are associated with the orientation, it is appropriate to model the bone as a transversely isotropic material. In this section, the finite element model of a femur with anisotropic materials in ANSYS is implemented.

4.2.1 FINITE ELEMENT MODEL OF FEMUR WITH ANISOTROPIC MATERIALS

Wolff's trajectorial theory states that the inner architecture of bone adapts to external influences. It indicates a remarkable similarity between the trabecular architecture of proximal femur and the stress trajectories. Therefore, the material orientation matches well with the principal stress track. The directions of the femoral neck and stem can be determined by the stress track in the femur.

Thus, an approach was proposed to assign the material directions from the stress field of the isotropic model (Figure 4.1). It involves the steps that are described next.

1. *Simulating the mechanical testing of a cadaver femur.* A finite element model of a man's femur was studied here (Figure 4.2), which comprises 42,217 elements and 124,954 nodes. With this model, the mechanical testing of a cadaver femur in a vertical (axial) loading was simulated in ANSYS, assuming that the bone was isotropic (Figure 4.3).

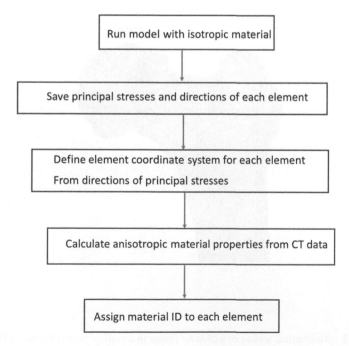

FIGURE 4.1 Flowchart for building the anisotropic finite element model of a femur.

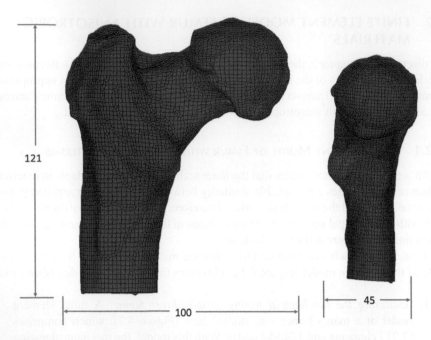

FIGURE 4.2   Finite element model of a femur (all dimensions in millimeters).

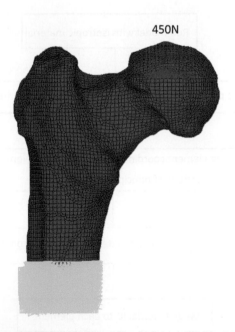

FIGURE 4.3   Mechanical testing of a cadaver femur in a vertical (axial) loading. (The bottom (green) is fixed.)

2. *Saving the principal stresses and directions of each element from the results of the isotropic model.* After solving this, the ETABLE command was used to acquire the principal stresses. The principal directions were saved by the command *VFUN, DIRCOS:

```
/POST1
*GET, NUMELEM, ELEM, 0, COUNT
*DIM, ARRAYS, ARRAY, NUMELEM, 6, 1, , ,
*DIM, DIRC, ARRAY, NUMELEM, 9, 1, , ,
*DIM, SPRINC, ARRAY, NUMELEM, 3, 1, , ,
*DIM, DIRE, ARRAY, NUMELEM, 6, 1, , ,
SET, LAST
AVPRIN, 0, ,
ETABLE, S_1, S, 1
AVPRIN, 0, ,
ETABLE, S_2, S, 2
AVPRIN, 0, ,
ETABLE, S_3, S, 3

*VGET, SPRINC(1,1), ELEM, 1, ETAB,S_1, , ,2  !get principal
                                                stresses
*VGET, SPRINC(1,2), ELEM, 1, ETAB, S_2, , ,2
*VGET, SPRINC(1,3), ELEM, 1, ETAB, S_3, , ,2
*VFUN, DIRC(1,1), DIRCOS, ARRAYS(1,1)      ! get directions of
                                             principal stresses
```

3. *Defining the element coordinate system for each element.* The origin of one element coordinate system is centered in the element, and its directions follow the stress principal directions. Thus, each element has its own element coordinate system:

```
*DO, E, 1, NUMELEM, 1    ! assign principal directions to the
                           element coordinate system
ICOR = 2E5 + E
CSYS, 0
ESEL, S, ELEM, , E
NSLE, S
X_ = CENTRX(E)             ! element center
Y_ = CENTRY(E)
Z_ = CENTRZ(E)
CSYS, 4
```

```
AA = ABS(DIRE(E,1)) + ABS(DIRE(E,2)) + ABS(DIRE(E,3))

*IF, AA, GT, 0.1, THEN

! create working plane along the directions of the
principal stresses.

WPLANE, 1, X_, Y_, Z_, X_ + DIRE(E,1), Y_ + DIRE(E,2), Z_ +
DIRE(E,3), X_ + DIRE(E,4), Y_ + DIRE(E,5), Z_ + DIRE(E,6)

CSWPLA, ICOR, 0    ! create local coordinate system associated
                     with the working plane

EMODIF, E, ESYS, ICOR    ! update ESYS with the defined local
                           coordinate system

*ENDIF

*ENDDO
```

4. *Defining the anisotropic material properties of the bone.* The anisotropic material properties of cortical and cancellous bone are specified by Eqs. (4.5) and (4.6), respectively [1].

  Cortical bone:

$$E_1 = E_2 = 2314\rho^{1.57}, E_3 = 2065\,\rho^{3.09}$$

$$\nu_{12} = 0.25, \ \nu_{13} = \nu_{23} = 0.\,32 \tag{4.5}$$

$$G_{12} = \frac{E_1}{2(1+\nu_{12})}, \ G_{13} = G_{23} = 3300\,\text{MPa}$$

  Cancellous bone:

$$E_1 = E_2 = 1157\rho^{1.78}, E_3 = 1094\,\rho^{1.64}$$

$$\nu_{12} = 0.25, \ \nu_{13} = \nu_{23} = 0.32 \tag{4.6}$$

$$G_{12} = \frac{E_1}{2(1+\nu_{12})}, \ G_{13} = G_{23} = 110\,\text{MPa}$$

  In Eqs. (4.5) and (4.6), the density ρ can be converted from the Young's modulus in the old isotropic model [2]:

$$\rho = \left(\frac{E}{7607}\right)^{1/1.853} \tag{4.7}$$

5. *Updating the material identity numbers in the model.* Each material identity in the isotropic model has a new corresponding material identity, with the anisotropic material properties obtained in step 3. Then, elements with an old material identity are assigned the corresponding new material identity to obtain anisotropic material properties.

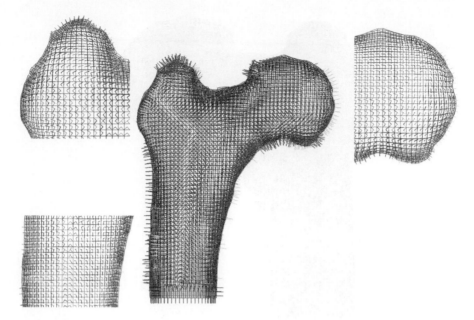

FIGURE 4.4 ESYSs at the cross section of a femur.

After these steps are finished, all the ESYSs in one cross section of the bone are plotted as shown in Figure 4.4. Obviously, ESYSs vary across the section and follow the stress principal directions.

4.2.2 SIMULATION OF MECHANICAL TESTING OF THE FEMUR

The developed finite element model was applied in order to simulate the mechanical testing of a cadaver femur in a vertical (axial) loading. The bottom was fixed in all degrees of freedom (DOFs), and a 450N force was applied on the bone in the axial loading (Figure 4.3).

The deformation and von Mises stresses are plotted in Figures 4.5 and 4.6. The axial loading bends the bone, with a maximum deformation of 0.76 mm. The main stress occurs at the end of the shaft, with a peak value of 12.3 MPa due to the bending moment that is caused by the axial loading.

4.2.3 DISCUSSION

The anisotropic model of a femur was created with a unique ESYS for each element. The ESYSs are determined by the principal stress directions obtained in the isotropic model. The defined ESYSs roughly match the directions of trabecular structures for the cancellous bone and the directions of the Haversian system for the cortical bone, which validates the proposed method.

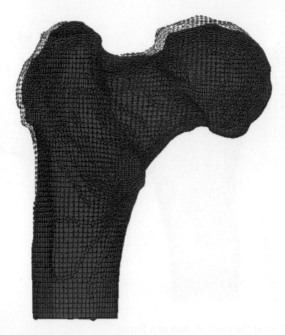

FIGURE 4.5 Deformation of the femur with anisotropic materials (with maximum deformation of 0.76 mm).

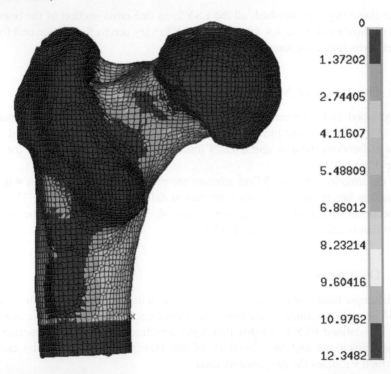

FIGURE 4.6 von Mises stresses of the femur with anisotropic materials (MPa).

4.2.4 SUMMARY

The anisotropic model of a femur was built in ANSYS, and each element has a unique ESYS determined by the principal stress direction from the isotropic model. The developed model can be used to simulate the mechanical testing of the bone.

REFERENCES

1. Wirtz, D. C., Schiffers, N., Pandorf, T., Radermacher, K., Weichert, D., and Forst, R., "Critical evaluation of known bone material properties to realize anisotropic FE-simulation of the proximal femur." *Journal of Biomechanics*, Vol. 33, 2000, pp. 1325–1330.
2. García, E. G., *Double Experimental Procedure for Model-Specific Finite Element Analysis of the Human Femur and Trabecular Bone*, Technische Universität München, 2013.

5 Simulation of Crack Growth Using the eXtended Finite Element Method (XFEM)

Chapter 5 aims to study cortical bone crack growth in the microstructure using the eXtended Finite Element Method (XFEM). The basics of XFEM are introduced in Section 5.1; and the modeling of bone crack growth is presented in Section 5.2.

5.1 INTRODUCTION TO XFEM

XFEM was developed to model both stationary crack problems and crack-growth simulation [1]. In crack growth, the crack front splits the existing elements without generating new elements. No morphing or remeshing is needed.

The following two methods classify XFEM:

1. If the crack terminates inside a finite element, it is called a *singularity-based method* (Figure 5.1a).
2. If the crack terminates only at the edge of a finite element, it is called a *phantom-node method* (Figure 5.1b).

These methods are described in the next subsections.

5.1.1 SINGULARITY-BASED METHOD

To capture the discontinuity in displacement across the crack surface and the crack-tip singularity, the displacement functions of the singularity-based method have enrichment functions other than the conventional shape functions [1]:

$$u(x) = N_I(x)u_I + H(x)N_I(x)a_I + N_I(x) \sum_j F_j(x)b_I^j, \qquad (5.1)$$

where $u(x)$ is the displacement vector, $N_I(x)$ is conventional nodal shape functions; u_I represents nodal displacement vectors; $H(x)$ is the Heaviside step function, defined by

$$H(x) = \begin{cases} 1, & x > 0 \\ 0, & x \leq 0 \end{cases}; \qquad (5.2)$$

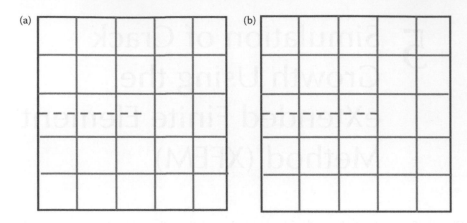

FIGURE 5.1   Schematic of XFEM. (a) Singularity-based method; (b) phantom-node method.

a_I is enriched nodal degrees of freedom (DOFs) for the displacement jump; $F_j(x)$ represents crack-tip functions; and b_I^j is nodal DOFs for the crack-tip singularity.

5.1.2   PHANTOM-NODE METHOD

As opposed to the singularity-based method, the phantom-node method ignores the crack-tip singularity contributions [1]. Thus, its displacement formulation is written as

$$u(x) = N_I(x)u_I + H(x)N_I(x)a_I. \qquad (5.3)$$

A superposed element formulation is introduced in order to split the parent element into two subelements, s1 and s2 (Figure 5.2). Thus, the displacement function

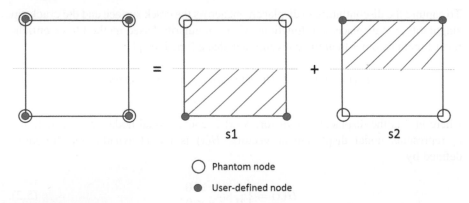

○   Phantom node

●   User-defined node

FIGURE 5.2   Schematic of a parent element divided into two subelements, s1 and s2.

is expressed in terms of displacement of the real nodes and the phantom nodes, as follows:

$$u(X, t) = u_I^{s1}(t)N_I(X)H(-f(X)) + u_I^{s2}(t)N_I(X)H(f(X)), \qquad (5.4)$$

where $u_I^{s1}(t)$ is the displacement vector in subelement s1; $u_I^{s2}(t)$ is the displacement vector in subelement s2; and $f(X)$ is the crack-surface definition ($f(X) = 0$).

5.1.3 GENERAL PROCESS FOR PERFORMING XFEM CRACK-GROWTH SIMULATION

The general process to perform XFEM-based crack-growth simulation includes the following steps [1]:

1. Define an initial crack.

   MESH200 is used to define the crack front, which is independent of the base mesh.

2. Specify the crack-growth criteria.

   Crack-growth criteria are specified as follows:

   ```
   TB, CGCR, ,,, STTMAX (or PSMAX)
   TBDATA, 1, VALUE
   ```

3. Calculate each crack-growth criterion.

   It involves the following definitions:

   ```
   CINT, CXFE, CompName       ! define the elements to form the crack front
   CINT, RADIUS, VALUE        ! define the distance ahead of the crack front
   ! define the parameters for the sweep angle
   CINT, RSWEEP, NUM_INTERVALS, MIN_ANGLE, MAX_ANGLE
   CINT, TYPE, STTMAX (or PSMAX)   ! calculate the circumferential stress
   ```

4. Perform the crack-growth calculation.

   Once solutions reach convergence, the crack-growth calculation is associated with the following commands:

   ```
   CGROW, NEW, SET_NUM           ! initiate the crack growth set
   CGROW, CID, ID_NUM            ! specify the crack calculation ID
   CGROW, METHOD, XFEM           ! specify xfem crack-growth method
   CGROW, FCOPTION, MTAB, MAT_ID ! specify the fracture criteria
   ```

5.2 SIMULATION OF CRACK GROWTH OF THE CORTICAL BONE

Osteoporosis (meaning "porous bone") (Figure 5.3) is a disease in which the density and quality of bone are reduced, leading to bone fragility and increased risk of fracture. Thus, it is of great importance to understand the damage and fracture of bone.

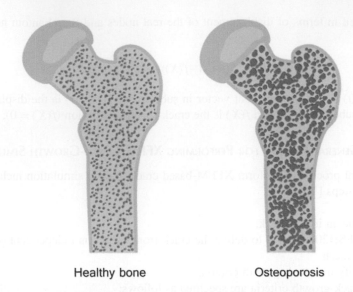

Healthy bone   Osteoporosis

FIGURE 5.3   Schematic of osteoporosis, as opposed to healthy bone. (Fancytapis © 123RF.com.)

Crack growth, rather than crack initiation, plays an active role in the bone-toughening mechanisms. The major sources of microscopic and macroscopic bone toughness come from the interaction of microcracks within the bone microstructure (Figure 5.4) to influence the crack paths through the tissue [2–3]. Normally, bone-crack propagation starts at the formation of a microcrack zone and then develops into a wake. Understanding this phenomenon will not only explain why the cortical bone is so resistant to crack propagation, but it also will provide a guideline to developing new biomaterials based on a biomimetic approach. Therefore, some computational models were built to investigate microcrack interactions with the bone microstructure [4–8]. In this instance, the crack growth was modeled by XFEM in ANSYS190.

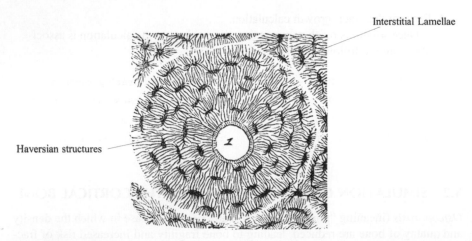

Interstitial Lamellae

Haversian structures

FIGURE 5.4   Microstructure of an osteon in cortical bone. (Patrick Guenette © 123RF.com.)

5.2.1 Finite Element Model

5.2.1.1 Geometry and Mesh

The finite element model has two osteons and one inclined crack between the two osteons (Figure 5.5). Each osteon was modeled as a circle with a radius of 0.100 mm. In its center is the Haversian channel, which is also modeled as a circle with a radius of 0.03 mm. Between the osteon and the interstitial matrix is the cement line, with a constant thickness of 0.003 mm.

The model was meshed with PLANE182 with key option (3) = 2 to represent plane strain conditions. The area around the crack was meshed regularly, and the other was meshed freely.

5.2.1.2 Material Properties

The osteon, the interstitial lamellae, and the cement are regarded as being linearly elastic. Their material properties are listed in Table 5.1 [8].

The maximum circumferential stress criterion (STTMAX) was selected for crack-growth criterion, as follows:

```
! crack-growth criterion
TB, CGCR, 2, , , STTMAX
TBDATA, 1, 10.5
```

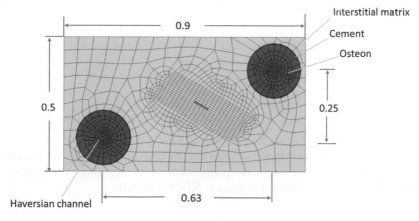

FIGURE 5.5 Finite element model of bone with an inclined crack (all dimensions in millimeters).

TABLE 5.1

Material Properties of the Cortical Bone Microstructure

	Young's Modulus (MPa)	Poisson's Ratio
Interstitial	14,000	0.15
Osteon	9,000	0.17
Cement	7,000	0.49

The decay of the cohesive stresses is governed by a rigid, linear, cohesive law defined by

```
! interface behavior
TB, CGCR, 2, , , RLIN
TBDATA, 1, 0.04, , 0.2
```

5.2.1.3 Definition of Crack Front

Two key points were created to form a line representing the crack front. The crack front was meshed by MESH200. Thus, the element component and node component of MESH200 were defined:

```
! mesh the crack with Mesh200 elements
ET, 2, 200, 0                              ! line with 2 key points
K, 110, (0.48170 + 0.48587)/2, (0.25575 + 0.26297)/2
K, 111, (0.53119 + 0.53536)/2, (0.22718 + 0.23440)/2
L, 110, 111
TYPE, 2
LMESH, 33
! element component for MESH200 elements
ESEL, S, TYPE, , 2
CM, M200EL, ELEM
! node component for crack front
ESEL, S, TYPE, , 2
NSLE, S
NSEL, U, LOC, X, 0.49, 0.52
CM, M200ND, NODE
```

The Phantom-node method was chosen for XFEM calculation. Thus, the crack-front data, including the element component and the node component of MESH200, were specified by the command XFCRKMESH:

```
! define enrichment identification
XFENRICH, ENRICH1, TESTCMP              ! for Phantom-node method.
XFCRKMESH, ENRICH1, M200EL, M200ND    ! define crack data
```

5.2.1.4 Local Coordinate Systems

The crack extensions of the two ends of the crack front are opposite each other and are not aligned with the global coordinate system. Therefore, two local coordinate systems were created by the CSWPLA command for them:

```
WPSTYLE, , , , , , , , 1
WPROTA, 150, ,                ! local coordinate 12 by rotating 150°
```

```
CSWPLA, 12, 0
WPROTA, -150
WPSTYLE, , , , , , , , 1
WPROTA, -30,,              ! local coordinate 11 by rotating -30°
CSWPLA, 11, 0
CSYS, 0
```

5.2.1.5 Loading and Boundary Conditions

The top edge applied negative pressure to simulate tension, and the left and bottom edges were constrained in the x- and y-directions, respectively.

5.2.1.6 Solution Setting

The inclined crack has two crack ends. Thus, the two ends have different definitions for fracture parameter calculation and crack-growth calculation:

```
! CINT calculations
CINT, NEW, 1
CINT, CXFE, _XFCRKFREL1
CINT, TYPE, STTMAX
CINT, NCON, 6
CINT, NORM, 12, 2
CINT, RSWEEP, 181, -90, 90
! CGROW calculations
CGROW, NEW, 1
CGROW, CID, 1
CGROW, METHOD, XFEM
CGROW, FCOPTION, MTAB, 2

! CINT calculations
CINT, NEW, 2
CINT, CXFE, _XFCRKFREL2
CINT, TYPE, STTMAX
CINT, NCON, 6
CINT, NORM,11, 2
CINT, RSWEEP, 181, -90, 90
! CGROW calculations
CGROW, NEW, 2
CGROW, CID, 2
CGROW, METHOD, XFEM
CGROW, FCOPTION, MTAB, 2
XFPR,1
```

5.2.2 RESULTS

Figure 5.6 illustrates the crack path, which indicates that the crack grows perpendicular to the loading direction and stops at the boundary of the enrichment area. The deformation and von Mises stresses are plotted in Figures 5.7 and 5.8. The deformation of the whole model is asymmetric because the structure is asymmetric. Some major stresses occur at the boundary of the osteons, where the material properties are not continuous. Table 5.2 lists the stress intensity factor K value of the two ends of a crack, which shows that the ends are path independent. The K values of the ends are 2.26 and 2.38, respectively.

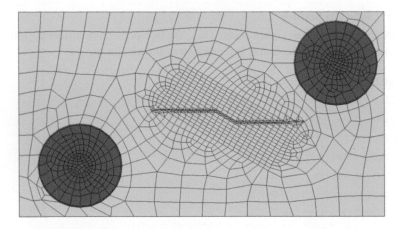

FIGURE 5.6 Crack growth of the inclined crack.

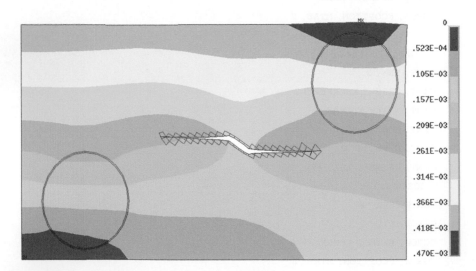

FIGURE 5.7 Deformation of the cortical bone (mm).

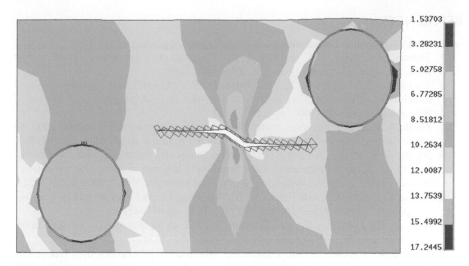

FIGURE 5.8   von Mises stresses of the cortical bone (MPa).

5.2.3   DISCUSSION

The crack growth between the two osteons was simulated with XFEM in ANSYS, which indicates that the softer osteons could attract the propagating crack. In addition, the stress intensity factor K of the crack front shows path independence. All these results demonstrate that the built model can be utilized to analyze the microcrack interactions with the bone microstructure.

XFEM has some advantages for simulating crack growth. Its crack front can be created by MESH200, independent of the base meshing, and the crack growth splits the elements at the crack front without the new meshing. A significant drawback of XFEM is that it requires regular meshing at the crack front, which does not exist in most practical problems.

5.2.4   SUMMARY

XFEM was applied to simulate the micro-crack growth of cortical bone in ANSYS. The constructed finite element model can be used to study the microcrack interactions within the bone microstructure.

TABLE 5.2
K_I of the Crack Front at Various Contours

Contour No.	1	2	3	4	5	6
Crack_id 1	1.98	2.25	2.26	2.27	2.27	2.26
Crack_id 2	2.21	2.39	2.38	2.38	2.38	2.39

REFERENCES

1. ANSYS 19.0 Help Documentation in the help page of product ANSYS190.
2. Mohsin, S., O'Brien, F. J., and Lee, T. C., "Osteonal crack barriers in ovine compact bone." *Journal of Anatomy*, Vol. 208, 2006, pp. 81–89.
3. Vashishth, D., Tanner, K. E., and Bonfield, W., "Contribution, development and morphology of microcracking in cortical bone during crack propagation." *Journal of Biomechanics*, Vol. 33, 2000, pp. 1169–1174.
4. Najafi, A. R. et al., "Haversian cortical bone model with many radial microcracks: An elastic analytic solution." *Medical Engineering & Physics*, Vol. 29, 2007, pp. 708–717.
5. Najafi, A. R. et al., "A fiber-ceramic matrix composite material model for osteonal cortical bone fracture micromechanics: Solution of arbitrary microcracks interaction." *Journal of the Mechanical Behavior of Biomedical Materials*, Vol. 2, 2009, pp. 217–223.
6. Huang, J., Rapoff, A. J., and Haftka, R. T., "Attracting cracks for arrestment in bone-like composites." *Materials and Design*, Vol. 27, 2006, pp. 461–469.
7. Vergani, L., Colombo, C., and Libonati, F., "Crack propagation in cortical bone: A numerical study." *Procedia Materials Science*, Vol. 3, 2014, pp. 1524–1529.
8. Baptista, R., Almeida, A., and Infante, V., "Micro-crack propagation on a biomimetic bone like composite material studied with the extended finite element method." *Procedia Structural Integrity*, Vol. 1, 2016, pp. 18–25.

Part II

Soft Tissues

Soft tissue comprises the tissues that connect, support, or enclose other structures and organs of the body, including tendons, ligaments, skin, fascia, fibrous tissues, muscles, nerves, and blood vessels. Some common diseases are linked with soft tissues. Abdominal aortic aneurysm (AAA) disease is due to the rupture of AAA wall tissue, and low back pain is related to mechanical behavior of the annulus and nucleus in the intervertebral disc. Therefore, Part II discusses finite element modeling of soft tissues.

Chapter 6 presented the structure and material properties of soft tissues. Then, Chapter 7 studied the nonlinear behavior of soft tissues and applied it to simulate AAA. The viscoelasticity of soft tissues and its application to the study of periodontal ligament creep were examined in Chapter 8. Chapter 9 described three ways to model fiber enhancement in ANSYS, while Chapter 10 introduced how to develop user material modes in ANSYS. Modeling soft tissues as porous media and its application in head impact and IVD creep study were discussed in Chapter 11.

6 Structure and Material Properties of Soft Tissues

Soft tissues exist throughout the human body. The structure and material properties of several common soft tissues, such as cartilage, ligaments, and intervertebral discs (IVDs), are introduced in this chapter in order to provide a better understanding of their mechanical behaviors in the chapters that follow.

6.1 CARTILAGE

6.1.1 STRUCTURE OF CARTILAGE

Articular cartilage is made of a multiphasic material containing fluid and solid elements. The fluid element is composed of water (68%–85%) with electrolytes, and the solid element consists of type II collagen fibrils (10%–20%), proteoglycans and other glycoproteins (5%–10%), and chondrocytes (cartilaginous cells). A total of 30% of all cartilage water exists in the interstitial fluid. When the cartilage is under compression, the fluid can flow from the tissue to the solution around the tissue; when the external loading is moved away, the fluid flows from the solution back to the tissue. The internal friction between the fluid and the tissue determines cartilage-compressive viscoelastic behaviors and provides the mechanism for energy dissipation.

The structure of cartilage is not uniform, but instead can be divided into four zones along its depth: superficial, middle, deep, and calcified (Figure 6.1) [1]. Each zone has its own biochemical, structural, and cellular characteristics. Flattened chondrocytes, low quantities of proteoglycan, and high quantities of collagen fibrils laid down parallel to the articular surface make up the superficial zone. The middle zone has round chondrocytes, high quantities of proteoglycan, and collagen in a random arrangement. Collagen fibrils perpendicular to the underlying bone and columns of chondrocytes arranged in the direction of the fibril axis characterize the deep zone. Finally, the calcified zone is a transition between the cartilage and the underlying subchondral bone, and partly mineralized. The distinct four zones along the cartilage depth with different structures make cartilage an anisotropic material.

6.1.2 MATERIAL PROPERTIES OF CARTILAGE

The contact in the joint always occurs among the cartilages. Interactions among the fluid, proteoglycan molecules, and various electrostatic charges act to provide superior lubrication and shock absorption. These interactions also make the coefficient of friction of cartilage extremely low, around 0.002 [2]. Such a low coefficient of friction guarantees the joint working well in a lifetime.

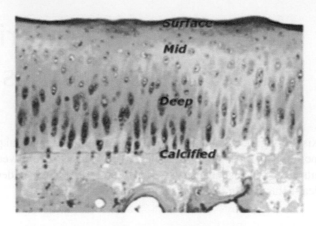

FIGURE 6.1   Cartilage structure (under the terms of the Creative Commons Attribution 3.0 License). (From Jun, H. et al., *Advances in Mechanical Engineering*, Vol. 7, 2015 [1].)

Like bone, cartilage is an anisotropic material, partly due to its structural variations. The cartilage is porous, which allows fluid to move in and out of the tissue. The mechanical properties of cartilage vary with its fluid content. Thus, it is very important to connect with the stress-strain history of the tissue to predict its load-carrying capacity.

6.2   LIGAMENTS

6.2.1   Structure of Ligaments

The ligaments are composed of collagen, elastin, glycoproteins, protein polysaccharides, glycolipids, water, and cells. Collagen and ground substances are the major components; their content and organization primarily determine the physical behavior of ligaments.

Water constitutes around two-thirds of the wet weight of ligaments. Collagen (especially type I) makes up 70%–80% of the dry weight of the ligament, while the ground substance comprises only around 1% of the ligament's dry weight. A large amount of the water exists in the ground substance, which is a source of the ligament's viscoelastic behavior.

Collagen fibers are packed closely and arranged in orientation [3], which provides motion and stability for the musculoskeletal system. Properties vary with strain rate, temperature, hydration, maturation, aging, immobilization, exercise, and healing.

6.2.2   Material Properties of Ligaments

A typical force-elongation curve of a ligament is shown in Figure 6.2 [4,5], which indicates that it is nearly linear before it fails the tensile test.

Ligaments have significant time-dependent and history-dependent viscoelastic properties. Time-dependent behavior indicates that a variety of load conditions during

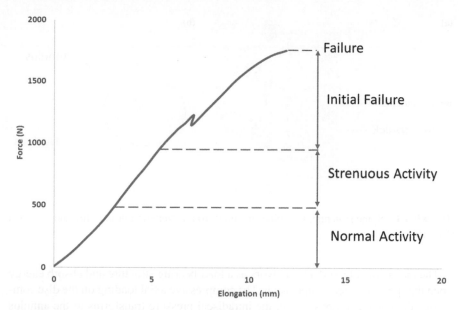

FIGURE 6.2 Force-elongation diagram of ligaments during tensile test. (From Pal, S., *Design of Artificial Human Joints & Organs*. Springer, 2014 [5].)

daily activities affect the mechanical properties of ligaments. On the other hand, history-dependent behavior means that frequent intense activities affect the tissue properties on a medium-term basis. In addition, ligaments are also sensitive to temperature change. The peak stresses of ligaments increase with decreased temperature.

The "crimp pattern" allows ligaments to have a certain range of strains with minimal resistance to movement [3]. Consequently, joints may move easily in certain ranges and directions. Once a joint is moved toward a range beyond the normal range of motion, the strain in the ligaments of the joint increases to make collagen fibers from their "crimped" state to a straightened condition; this significantly increases the resistance of ligaments to further elongation.

6.3 INTERVERTEBRAL DISC

Vertebral discs are the shock-absorbing structures in the spine (Figure 6.3a). Discs have identical structures, although they vary in size and shape depending on the region of the spine. The outside of the discs is enclosed by collagenous annulus fibrosus, and its central part is the nucleus (Figure 6.3b). The inferior and superior end plates of the adjacent vertebra are located on the top and bottom surfaces of the discs.

The annulus and nucleus are both multiphasic. The annulus fibrosus has water ranging from 65% to 70%. Collagen in the annulus accounts for about half of its dry weight. The nucleus has a much higher water content than the annulus (as high as 88%), but it also has much less collagen (mostly type II), amounting to about 20%–30% of the dry weight. Most of the material in the nucleus is composed of proteoglycan and other proteinaceous materials.

(a)   (b)

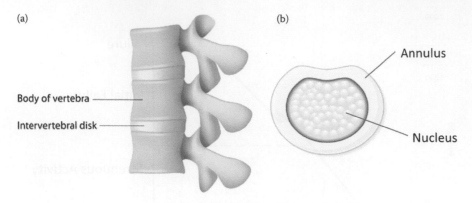

Annulus

Body of vertebra

Intervertebral disk

Nucleus

FIGURE 6.3   Spine structure. (a) Spine structure; (b) intervertebral disc. (Alila and designua © 123RF.com.)

The major role of the IVD is to work as a load-bearing structure and absorb energy when the spine is under compression. The compressive axial loading on the disc compresses the nucleus pulposus, and the intradiscal pressure transforms to the annulus fibrosus, causing it to swell in the radical direction like "barreling" [6]. The annulus absorbs most of the barreling of the disc via collagen network elongation. In addition, a large amount of viscous water in the disc allows it to act as a hydrostatic shock absorber.

REFERENCES

1. Jun, H., Evans, T. M., and Mente, P. L., "Study on the damage mechanism of articular cartilage based on the fluid-solid coupled particle model." *Advances in Mechanical Engineering*, Vol. 7, 2015.
2. Guilak, F., "The slippery slope of arthritis." *Arthritis Rheumatology*, Vol. 52, 2005, pp. 1632–1633.
3. Kastelic, J., Palley, I., and Baer, E., "The multicomposite ultrastructure of tendon." *Connective Tissue Research*, Vol. 6, 1978, pp. 11–23.
4. Fung, Y. C., *Biomechanics: Mechanical Properties of Living Tissues*. 2nd ed., Springer, New York, 1993.
5. Pal, S., *Design of Artificial Human Joints & Organs*. Springer, New York, 2014.
6. Koopman, W. J. and Moreland, L. W., *Arthritis and Allied Conditions*. 15th ed., Lippincott Williams & Wilkins, Philadelphia, 2005.

7 Nonlinear Behavior of Soft Tissues

As mentioned in Chapter 6, soft tissues exhibit a strongly nonlinear strain-stress relation, which can be modeled as hyperelastic materials. In this chapter, after summarizing the hyperelastic models in Section 7.1, a finite element analysis of abdominal aortic aneurysm wall is conducted in Section 7.2.

7.1 HYPERELASTIC MODELS

Nomenclature

The terms used in this discussion are as follows:

W	Strain energy for each unit of volume
S_{ij}	Components of the second Piola-Kirchhoff stress tensor
τ_{ij}	Components of the Kirchhoff stress
σ_{ij}	Components of the Cauchy stress
E_{ij}	Components of the Lagrangian strain tensor
C_{ij}	Components of the right Cauchy-Green deformation tensor
δ_{ij}	Kronecker delta
F_{ij}	Components of the deformation gradient tensor
I_1	First deviatoric strain invariant
I_2	Second deviatoric strain invariant
I_3	Third deviatoric strain invariant
J	Determinant of the elastic deformation gradient
J_m	Limiting value of $I_1 - 3$
μ	Initial shear modulus
λ_L	Limiting network stretch
$\bar{\lambda}_p$ $(p=1,2,3)$	Deviatoric principal stretches
d	Material incompressibility parameter

Hyperelastic materials are characterized by a relatively low elastic modulus and high bulk modulus. They are commonly subjected to large strains and deformations. Generally, they are defined by an elastic potential W (or strain energy density function), which is a scalar function of one of the strain or deformation tensors. Their stress components are determined by the derivative of W with respect to a strain component. The second Piola-Kirchhoff stress tensor can be expressed by

$$S_{ij} = \frac{\partial W}{\partial E_{ij}} = 2\frac{\partial W}{\partial C_{ij}}. \qquad (7.1)$$

The Lagrangian strain takes the following form:

$$E_{ij} = \frac{1}{2}(C_{ij} - \delta_{ij})$$
(7.2)

The deformation tensor C_{ij} is determined by the deformation gradients F_{ij}:

$$C_{ij} = F_{ki}F_{kj}.$$
(7.3)

The Kirchhoff stress is obtained by

$$\tau_{ij} = F_{ik}S_{kl}F_{jl}.$$
(7.4)

The Cauchy stress is expressed by

$$\sigma_{ij} = \frac{1}{J}F_{ik}S_{kl}F_{jl}.$$
(7.5)

Some hyperelastic material models with different strain energy density functions are available in ANSYS [1], and they are summarized as follows:

Arruda–Boyce Hyperelasticity

$$W = \mu\left[\frac{1}{2}(I_1 - 3) + \frac{1}{20\lambda_L^2}(I_1^2 - 9) + \frac{11}{1050\lambda_L^4}(I_1^3 - 27) + \frac{19}{7000\lambda_L^6}(I_1^4 - 81)\right.$$

$$\left. + \frac{519}{673750\lambda_L^8}(I_1^5 - 243)\right] + \frac{1}{d}\left(\frac{J^2 - 1}{2} - \ln J\right)$$
(7.6)

Blatz-Ko Foam Hyperelasticity

$$W = \frac{\mu}{2}\left(\frac{I_2}{I_3} + 2\sqrt{I_3} - 5\right)$$
(7.7)

Gent Hyperelasticity

$$W = -\frac{\mu J_m}{2}\ln\left(1 - \frac{I_1 - 3}{J_m}\right) + \frac{1}{d}\left(\frac{J^2 - 1}{2} - \ln J\right)$$
(7.8)

Mooney-Rivlin Hyperelasticity

Mooney-Rivlin hyperelasticity takes various forms with different parameters. Here is nine-parameter Mooney-Rivlin:

$$W = c_{10}(I_1 - 3) + c_{01}(I_2 - 3) + c_{20}(I_1 - 3)^2 + c_{11}(I_1 - 3)(I_2 - 3) + c_{02}(I_2 - 3)^2$$

$$+ c_{30}(I_1 - 3)^3 + c_{21}(I_1 - 3)^2(I_2 - 3) + c_{12}(I_1 - 3)(I_2 - 3)^2$$

$$+ c_{03}(I_2 - 3)^3 + \frac{1}{d}(J - 1)^2$$
(7.9)

TABLE 7.1
Features of Various Hyperelastic Materials

Hyperelastic Material Model	Material for Which the Model Works	Compressibility	Maximum Applicable Strain Level
Arruda-Boyce	Rubbers such as silicon and neoprene	Nearly incompressible	300%
Blatz-Ko Foam	Compressible polyurethane foam	Compressible	
Gent Hyperelasticity	Rubber	Nearly incompressible	300%
Mooney-Rivlin	Rubber for tires	Nearly incompressible	200%
Neo-Hookean		Nearly incompressible	20%–30%
Ogden	Material with large strain	Nearly incompressible	700%
Polynomial Form		Nearly incompressible	Almost the same as Mooney-Rivlin
Yeoh		Nearly incompressible	

Neo-Hookean Hyperelasticity

$$W = \frac{\mu}{2}(I_1 - 3) + \frac{1}{d}(J - 1)^2 \tag{7.10}$$

Ogden Hyperelasticity

$$W = \sum_{i=1}^{N} \frac{\mu_i}{\alpha_i} \left(\bar{\lambda}_1^{\alpha_i} + \bar{\lambda}_2^{\alpha_i} + \bar{\lambda}_3^{\alpha_i} - 3 \right) + \sum_{k}^{N} \frac{1}{d_k}(J - 1)^{2k} \tag{7.11}$$

Polynomial Form Hyperelasticity

$$W = \sum_{i+j=1}^{N} c_{ij}(I_1 - 3)^i (I_2 - 3)^j + \sum_{k}^{N} \frac{1}{d_k}(J - 1)^{2k} \tag{7.12}$$

Yeoh Hyperelasticity

$$W = \sum_{i=1}^{N} c_{i0}(I_1 - 3)^i + \sum_{k}^{N} \frac{1}{d_k}(J - 1)^{2k} \tag{7.13}$$

Table 7.1 lists the unique features of these hyperelastic materials.

7.2 FINITE ELEMENT ANALYSIS OF THE ABDOMINAL AORTIC ANEURYSM WALL

About 2 million Americans suffer from abdominal aortic aneurysm (AAA) disease, and the rate of occurrence is increasing. Without treatment, the aneurysm diameter enlarges at the rate of about 0.4 cm/year until rupture (Figure 7.1). Rupture of an

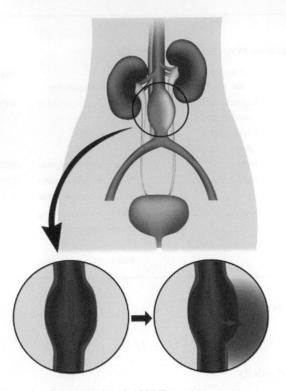

FIGURE 7.1   Schematic of AAA. (Alila © 123RF.com.)

AAA ranks thirteenth on the list of the most common causes of death in the United States [2]. Current clinical management of AAA patients uses the maximum transverse dimension of the aneurysmis (5 cm) as the primary indicator of the potential for rupture. However, it was found that some AAAs ruptured at a size < 5 cm, and other AAAs enlarged to 8 cm without rupturing. Thus, some studies suggest that AAA rupture occurs when the mechanical stress of the wall exceeds the strength of the wall tissue. Therefore, knowledge of the mechanical stresses in an intact AAA wall could be useful to assess the wall's risk of rupture. The finite element method was applied to study the wall stress of AAA intensively [3–13]. In this section, the finite element model of AAA was built in ANSYS.

7.2.1 Finite Element Model

7.2.1.1 Geometry and Mesh

A half-aneurysm model was generated with the computer-aided design (CAD) software program SpaceClaim, with a 2-cm diameter at the inlet and outlet sections and a maximum diameter of 6 cm at the midsection of the AAA sac (Figure 7.2a). The asymmetry of the model is governed by the β parameter, defined by [9]

$$\beta = \frac{r}{R}, \tag{7.14}$$

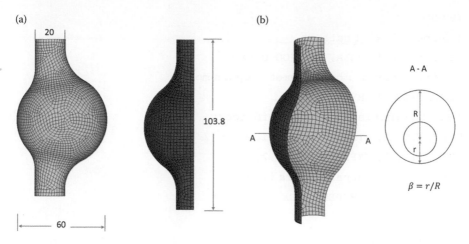

FIGURE 7.2 Finite element model of AAA. (a) Finite element model of AAA (all dimensions in millimeters); (b) asymmetric definition of AAA.

where r and R are the radii measured at the midsection of the AAA sac (shown in Figure 7.2b).

In this model, β was selected as 0.6.

The AAA wall was assumed to have a uniform thickness of 1.5 mm. The whole model, which was meshed with SHELL181, comprises 3,574 elements and 5,500 nodes.

7.2.1.2 Material Model

Curve-fitting of the material parameters was performed with experimental data. Experiments show that the aneurysmal tissue is nonlinear and undergoes large strains of up to 40% prior to failure [14]. Therefore, the Ogden hyperelastic model was selected for curve-fitting. The curve-fitting of AAA material parameters was conducted using the following ANSYS Parametric Design Language (APDL) commands:

```
*CREATE, CFHY-OG-U.EXP    ! input the experimental data
! engineering strain      engineering stress (MPa)
0.020414182               0.008333333
0.026507186               0.013333333
0.034005163               0.025416667
0.044289499               0.041666667
0.052914867               0.05625
0.060533894               0.075
0.065539987               0.0875
0.069668097               0.1
*END
```

```
/PREP7
! define uniaxial data
TBFT, EADD, 1, UNIA, CFHY-OG-U.EXP
! define material: one parameter Ogden model
TBFT, FADD, 1, HYPER, OGDE, 1
! set the initial value of coefficients
TBFT, SET, 1, HYPER, OGDE, 1, 1, 1
TBFT, SET, 1, HYPER, OGDE, 1, 2, 1
! define solution parameters if any
TBFT, SOLVE, 1, HYPER, OGDE, 1, 1, 500
/OUT

! print the results
TBFT, LIST, 1
```

The last command shown here prints out the Ogden parameters after the solution. Figure 7.3 presents curve-fitting results that match the experimental data well. Therefore, the material model was defined by

```
TB, HYPE, 1, 1, 1, OGDEN    ! one parameter Ogden model
TBDATA, 1, 0.005, 46.88
```

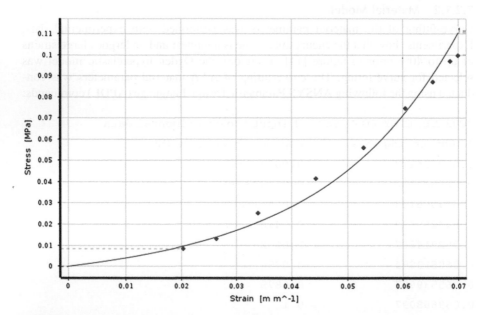

FIGURE 7.3 Curve-fitting results of an AAA material model.

7.2.1.3 Loading and Boundary Conditions

Using the SF command in ANSYS, 14,665 Pa (about 110 mmHg) was loaded on the surface of the AAA wall. The two ends should be constrained in the vertical direction but allow radial expansion without rigid body motion. Thus, force-distributed boundary constraints were applied to both sides of the artery (Figure 7.4). The following code presents the APDL commands for constraints on one end:

```
MAT, 4

R, 12

REAL, 12

ET, 11, 170

ET, 12, 175

KEYOPT, 12, 12, 5          ! bonded (always)

KEYOPT, 12, 4, 1           ! force distributed constraint

KEYOPT, 12, 2, 2           ! multipoint constraint

KEYOPT, 11, 2, 1

KEYOPT, 11, 4, 111111      ! all DOFs are constrained

TYPE, 12

NSEL, S, LOC, Y, -50

ESLN, S, 0

ESURF
```

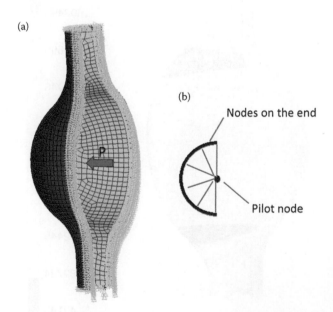

FIGURE 7.4 Boundary conditions of the AAA model. (a) Boundary condition of AAA model; (b) MPC constraint (pilot node coupled with nodes on the end).

```
TYPE, 11
! create a pilot node
N, 10004, -7.902, -50, 0
TSHAP, PILO
E, 10004
! generate the contact surface
D, 10004, ALL

ALLSEL
```

7.2.1.4 Solution Setting

A static analysis was conducted. Large deflection effects were turned on with the command NLGEOM, ON, as the hyperelastic model was selected for the AAA wall.

7.2.2 RESULTS

Figure 7.5 illustrates the swelling of the AAA wall under internal pressure of 110 mmHg. The maximum deformation is 5.43 mm. The deformation is asymmetric because the AAA is asymmetric with $\beta = 0.6$. Figure 7.6 plots the von Mises stresses

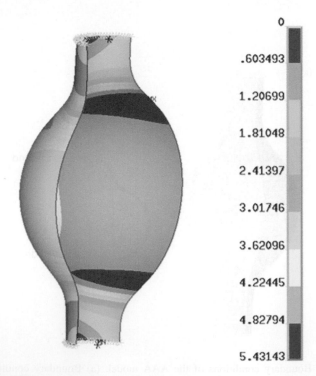

FIGURE 7.5 Deformation of the AAA (mm).

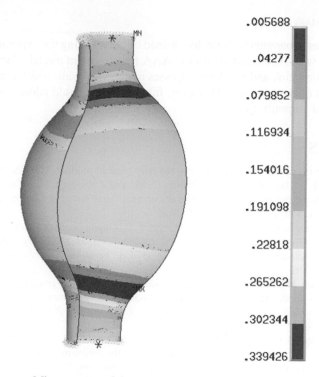

.005688	
.04277	
.079852	
.116934	
.154016	
.191098	
.22818	
.265262	
.302344	
.339426	

FIGURE 7.6   von Mises stresses of the AAA wall (MPa).

of the AAA wall. The maximum stress of 0.34 MPa occurs at the transition of the aorta and aneurysm where the geometry dramatically changes.

7.2.3   DISCUSSION

The AAA wall under compression was simulated in ANSYS with the Ogden hypere-lastic model. The obtained mechanical stresses could be used to assess the risk of AAA rupture. However, this work has some limitations. The first limitation is obvious: A smooth, idealized geometry was used instead of the complex geometry of a real AAA. However, the geometry of an AAA, including the AAA diameter and asymmetry of a real-life AAA, plays a significant role in determining the wall-stress distribution. Therefore, it is imperative to build a finite element model of an AAA based on the real AAA geometry. In addition, the uniform wall thickness of 1.5 mm was assumed in the model, which does not match the real AAA. Furthermore, the normal arterial tissue was demonstrated to be anisotropic. A more appropriate material model than hyperplastic isotropic materials should be selected for the AAA model.

Force-distributed boundary constraints were applied to the two ends of the AAA. Two ends of the pipe expand freely under compression, which is consistent with the real case. Simple constraints of all degrees of freedom (DOFs) at the ends may cause extra stress.

7.2.4 SUMMARY

The AAA material parameters were determined by curve-fitting the experimental data, and then they were incorporated into the AAA finite element model. The simulation output the deformation and von Mises stresses of the AAA wall, which can be used to assess the risk of AAA rupture. However, future studies should address some of the limitations that currently exist.

REFERENCES

1. ANSYS19.0 Help Documentation in the help page of product ANSYS190.
2. Patel, M. I., Hardman, D. T. A., Fisher, C. M., and Appleberg, M., "Current views on the pathogenesis of abdominal aortic aneurysms." *Journal of the American College of Surgeons*, Vol. 181, 1985, pp. 371–382.
3. Di Martino, E. S., Bohra, A., Vande Geest, J. P., Gupta, N., Makaroun, M., and Vorp, D. A., "Biomechanical properties of ruptured versus electively repaired abdominal aortic aneurysm wall tissue." *Journal of Vascular Surgery*, Vol. 43, 2006, pp. 570–576.
4. Raghavan, M., Kratzberg, J., and da Silva, E. S., "Heterogeneous, variable wall-thickness modeling of a ruptured abdominal aortic aneurysm." *Proceedings of the 2004 International Mechanical Engineering Congress and R&D Expo IMECE2004*, 2004.
5. Raghavan, M. L., Vorp, D. A., Federle, M. P., Makaroun, M. S., and Webster, M. W., "Wall stress distribution on three-dimensionally reconstructed models of human abdominal aortic aneurysm." *Journal of Vascular Surgery*, Vol. 31, 2000, pp. 760–769.
6. Vorp, D. A., Raghavan, M. L., and Webster, M. W., "Mechanical wall stress in abdominal aortic aneurysm: Influence of diameter and asymmetry." *Journal of Vascular Surgery*, Vol. 27, 1998, pp. 632–639.
7. Wilson, K. A., Lee, A. J., Hoskins, P. R., Fowkes, F. G. R., Ruckley, C. V., and Bradbury, A. W., "The relationship between aortic wall distensibility and rupture of infrarenal abdominal aortic aneurysm." *Journal of Vascular Surgery*, Vol. 37, 2003, pp. 112–117.
8. Raghavan, M. L., Kratzberg, J., Castro de Tolosa, E. M., Hanaoka, M. M., Walker, P., and da Silva, E. S., "Regional distribution of wall thickness and failure properties of human abdominal aortic aneurysm." *Journal of Biomechanics*, Vol. 39, 2006, pp. 3010–3016.
9. Scotti, C. M., Jimenez, J., Muluk, S. C., and Finol, E. A., "Wall stress and flow dynamics in abdominal aortic aneurysms: Finite element analysis vs. fluid-structure interaction." *Computer Methods in Biomechanics and Biomedical Engineering*, Vol. 11, 2008, pp. 301–322.
10. Scotti, C. M., Shkolnik, A. D., Muluk, S., and Finol, E. A., "Fluid–structure interaction in abdominal aortic aneurysms: Effects of asymmetry and wall thickness." *Biomedical Engineering Online*, Vol. 5, 2005, pp. 64.
11. Raut, S., Chandra, S., Shum, J., Washington, C. B., Muluk, S. C., Finol, E. A., and Rodriguez, J. F., "Biological, geometric, and biomechanical factors influencing abdominal aortic aneurysm rupture risk: A comprehensive review." *Recent Patents on Medical Imaging*, Vol. 3, 2013, pp. 44–59.
12. de Ruiz, G. S., Antón, R. S., Cazón, A., Larraona, G. S., and Finol, E. A., "Anisotropic abdominal aortic aneurysm replicas with biaxial material characterization." *Medical Engineering & Physics*, Vol. 38, 2016, pp. 1505–1512.

13. de Ruiz, G. S., Cazón, A., Antón, R., and Finol, E. A., "The relationship between surface curvature and abdominal aortic aneurysm wall stress." *Journal of Biomechanical Engineering*, Vol. 139, 2017, doi: 10.1115/1.4036826
14. Raghavan, M. L., Webster, M. W., and Vorp, D. A., "Ex-vivo biomechanical behavior of abdominal aortic aneurysm: Assessment using a new mathematical model." *Annals of Biomedical Engineering*, Vol. 24, 1996, pp. 573–582.

8 Viscoelasticity of Soft Tissues

As discussed in Chapter 6, soft tissues have significant time-dependent and viscoelastic properties. The Maxwell viscoelastic model is introduced in Section 8.1, and then it is used to study the creep test of the periodontal ligament (PDL) in Section 8.2.

8.1 THE MAXWELL MODEL

Viscoelasticity is the property of materials that is characterized by both viscous behavior and elastic behavior when undergoing deformation. This kind of material exhibits time-dependent strain. Some models, which include the Maxwell model, the Kelvin-Voigt model, and the Burgers model, have been developed to simulate viscoelasticity. These models are introduced briefly next.

The Maxwell model is composed of a purely viscous damper and a purely elastic spring connected in a series (Figure 8.1). It can be expressed as

$$\sigma + \frac{\eta}{E}\dot{\sigma} = \eta\dot{\varepsilon}, \tag{8.1}$$

where σ is the stress, E is the elastic modulus of the material, ε is the strain, and η is the viscosity of the material.

The Kelvin-Voigt model consists of a purely viscous damper and a purely elastic spring connected in parallel (Figure 8.2). It can be represented by

$$\sigma = E\varepsilon + \eta\dot{\varepsilon}. \tag{8.2}$$

The Burgers model is the combination of the Maxwell and Kelvin-Voigt models in series (Figure 8.3). It can be written as

$$\sigma + \left(\frac{\eta_1}{E_1} + \frac{\eta_2}{E_1} + \frac{\eta_2}{E_2}\right)\dot{\sigma} + \frac{\eta_1}{E_1}\frac{\eta_2}{E_2}\ddot{\sigma} = \eta_2\dot{\varepsilon} + \frac{\eta_1\eta_2}{E_1}\ddot{\varepsilon}. \tag{8.3}$$

The Maxwell model is widely used to simulate viscoelastic materials. A generalized Maxwell model in three dimensions is described as [1]

$$\sigma = \int_0^t 2G(t-\tau)\frac{d\mathbf{e}}{d\tau}d\tau + \mathbf{I}\int_0^t K(t-\tau)\frac{d\Delta}{d\tau}d\tau, \tag{8.4}$$

where σ is Cauchy stress; $\mathbf{e}$ is deviatoric strain; Δ is volumetric strain; τ is time; $\mathbf{I}$ is the identity tensor; $G(t)$ is the Prony series shear-relaxation moduli, which

61

FIGURE 8.1 Schematic of the Maxwell viscoelastic model. A viscous damper and a spring are connected in a series.

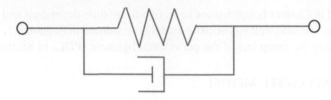

FIGURE 8.2 Schematic of the Kelvin-Voigt viscoelastic model. A viscous damper and a spring are connected in parallel.

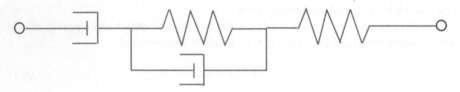

FIGURE 8.3 Schematic of the Burgers viscoelastic model. A Maxwell model and a Kelvin-Voigt model are combined in a series.

is determined by

$$G(t) = G_0 \left[\alpha_\infty^G + \sum_{i=1}^{n_g} \alpha_i^G \exp\left(-\frac{t}{\tau_i^G}\right) \right]; \tag{8.5}$$

$K(t)$ is the Prony series bulk-relaxation moduli:

$$K(t) = K_0 \left[\alpha_\infty^K + \sum_{i=1}^{n_k} \alpha_i^K \exp\left(-\frac{t}{\tau_i^K}\right) \right]; \tag{8.6}$$

G_0 and K_0 are the relaxation moduli at $t = 0$:

$$G_0 = \frac{E}{2(1+\nu)}, \; K_0 = \frac{E}{3(1-2\nu)};$$

n_g and n_k represent the number of Prony terms; α_i^G and α_i^K are the relative moduli; and τ_i^G and τ_i^K represent relaxation time.

In ANSYS, Prony series constants are defined by the following commands:

```
MP, EX, 1, E              ! define Young's modulus
MP, NUXY, 1, ν            ! define Poisson's ratio
TB, PRONY,,, NG, SHEAR    ! the shear Prony data table
TBDATA, 1, α₁ᴳ, τ₁ᴳ, ⋯ αₙᴳ, τₙᴳ,
TB, PRONY,,, NK, BULK     ! the bulk Prony data table
TBDATA, 1, α₁ᵏ, τ₁ᵏ, ⋯ αₙᵏ, τₙᵏ,
```

8.2 STUDY OF PDL CREEP

The PDL is a connective tissue that links the tooth root to the alveolus bone (Figure 8.4). It transfers orthodontic forces to the alveolar bone during orthodontic treatment. Orthodontic treatment is a long-term process, with accompanying possible side effects. Dentists are unclear about how much force should occur during dental surgery. To understand the role of PDL in transmitting treatment load, it is very important to study its material properties. In the past, many studies assumed that the PDL was linear [2], bilinear [3], and nonlinear [4] in finite element analysis. However, an experimental test indicates it may be more appropriate to model PDL as viscoelastic [5]. In this section, its creep test was simulated in ANSYS.

8.2.1 FINITE ELEMENT MODEL

8.2.1.1 Geometry and Mesh

A tooth finite element model was built in ANSYS [6]. The size and shape of the model, shown in Figure 8.5, were selected to be comparable to a human tooth. The tooth root was covered by a 0.2-mm-thick layer to represent the PDL. The tooth and PDL were enclosed by a block to represent the cancellous bone. A 2.5-mm-thick

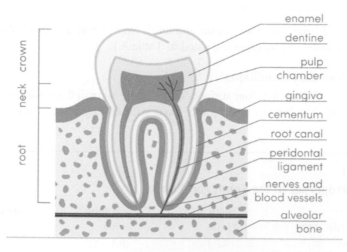

FIGURE 8.4 Dental structure. (masia8 © 123RF.com.)

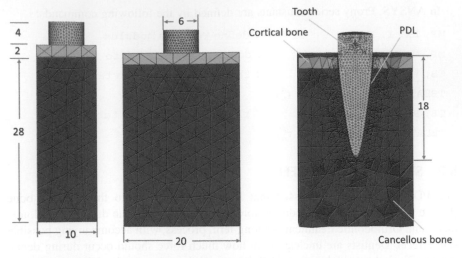

FIGURE 8.5 Finite element model of a tooth (all dimensions in millimeters).

layer of cortical bone sits on top of the cancellous bone. The whole model was meshed with SOLID187, comprised of 42,854 elements and 59,119 nodes.

8.2.1.2 Material Models

PDL was modeled by the three-parameter Prony series [7,8]:

```
MP, EX, 2, 0.23,                   ! MPa
MP, NUXY, 2, 0.49,
TB, PRONY, 2, ,3, SHEAR           ! define viscosity parameters
TBDATA, 1, 0.91, 0.0025, 0.05, 0.1, 0.005, 0.5
TB, PRONY, 2, ,3 ,BULK            ! define viscosity parameters
TBDATA, 1, 0.155, 0.0025, 0.4, 0.1, 0.15, 0.5
```

The cortical bone, cancellous bone, and tooth were assumed to be elastic and isotropic. Their material properties are listed in Table 8.1.

8.2.1.3 Boundary Conditions

To avoid overconstraining, four surfaces of the bones were defined with symmetrical conditions [6]. The bottom surface was fixed with all degrees of freedom (DOFs) (Figure 8.6).

TABLE 8.1

Material Properties of the Tooth and Bone

	Tooth	Cortical Bone	Cancellous Bone
Young's modulus (MPa)	13,700	13,700	1,370
Poisson's ratio	0.3	0.3	0.3

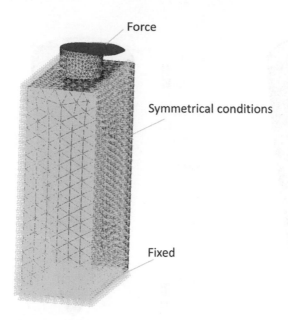

FIGURE 8.6 Boundary conditions of the tooth model.

8.2.1.4 Loading Steps

The simulation of the creep test [7] involved two loadings steps. A force of 1.45N was applied to the top surface of the tooth and held for 2.5 s in the first step. At the beginning of the second step, the loading was released. The whole creep test took 6 s.

8.2.2 RESULTS

The deformation and stresses of the PDL at 2.5 s are shown in Figure 8.7. As the load is loaded in one direction, the tooth shifts along the loading direction. Therefore, the top edge of the PDL has significant deformation and large stresses, with the maximum stress being 11 kPa. After the loading is released, the deformation and stresses of the PDL decay quickly (Figure 8.8). The maximum stress at 6 s is 0.204 kPa. Figure 8.9 illustrates the displacement time history of one node of the PDL, which has the exact curve shape as the experimental creep data [5].

8.2.3 DISCUSSION

The tooth finite element model was developed; it defined PDL by the three-parameter Prony series. Simulation of the creep test was conducted, and the results show that after the loading is released, the tooth gradually recovers, which is due to the viscoelasticity of the PDL. If the PDL is modeled as linearly elastic, the tooth will recover instantaneously.

One of the drawbacks of this study is that the model was simplified with an idealized geometry. A three-dimensional (3D), realistic model obtained from a magnetic

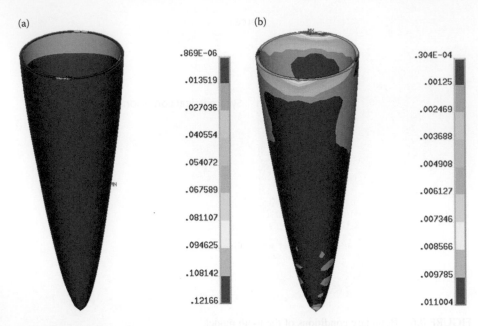

FIGURE 8.7 Deformation and von Mises stresses of PDL at 2.5 s. (a) Deformation (mm); (b) von Mises stresses (MPa).

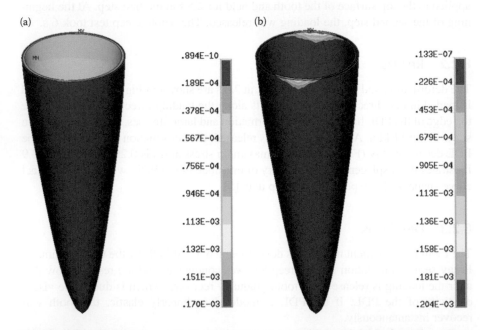

FIGURE 8.8 Deformation and von Mises stresses of PDL at 6s. (a) Deformation (mm); (b) von Mises stresses (MPa).

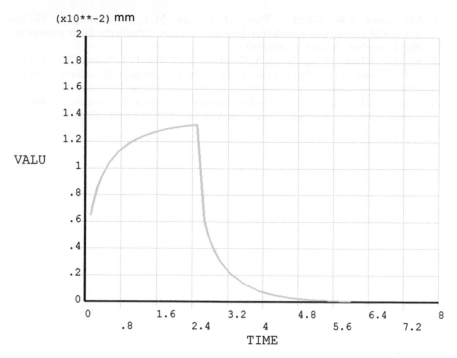

FIGURE 8.9 Time history of the displacement of node 47,680 in the z-direction.

resonance imaging (MRI) or computed tomography (CT) scan will produce more precise simulation results.

8.2.4 SUMMARY

In this chapter, the tooth PDL finite element model was created, and the PDL was modeled with the viscoelastic material. The creep test was simulated, and the results match the experimental data.

REFERENCES

1. ANSYS190 Help Documentation in the help page of product ANSYS190.
2. Motoyoshi, M., Hirabayashi, M., Shimazaki, T., and Namura, S., "An experimental study on mandibular expansion: Increases in arch width and perimeter." *European Journal of Orthodontics*, Vol. 24, 2002, pp. 125–130.
3. Katona, T. R. and Qian, H., "A mechanism of noncontinuous supraosseous tooth eruption." *American Journal of Orthodontics & Dentofacial Orthopedics*, Vol. 120, 2001, pp. 263–271.
4. Tanne, K., "Stress induced in the periodontal tissue at the initial phase of the application of various types of orthodontic forces: 3-dimensional analysis using a finite element method." *Osaka Daigaku Shigaku Zasshi*, Vol. 28, 1983, pp. 209–261.
5. Ross, G. G., Lear, C. S., and Decos, R., "Modeling the lateral movements of teeth." *Journal of Biomechanics*, Vol. 9, 1976, pp. 723–734.

6. McCormack, S. W., Witzel, U., Watson, P. J., Fagan, M. J., and Gröning, F., "The biomechanical function of periodontal ligament fibres in orthodontic tooth movement." *PLOS One*, Vol. 9, 2014, p. e102387.
7. Su, M. Z., Chang, H. H., Chiang, Y. C., Cheng, J. H., Fuh, L. J., Wang, C. Y., and Lin, C. P., "Modeling viscoelastic behavior of periodontal ligament with nonlinear finite element analysis." *Journal of Dental Sciences*, Vol. 8, 2013, pp. 121–128.
8. Yang, Y. and Tang, W., "Analysis of mechanical properties at different levels of the periodontal ligament." *Biomedical Research*, Vol. 28, 2017, pp. 8958–8965.

9 Fiber Enhancement

Fibers exist in some soft tissues to enhance their mechanical behavior. Some advanced finite element technologies have been developed to model fiber enhancement, such as standard fiber enhancement, mesh-independent fiber enhancement, and material models including fiber enhancement. These technologies are discussed in Sections 9.1 through 9.3, respectively.

9.1 STANDARD FIBER ENHANCEMENT

9.1.1 INTRODUCTION OF STANDARD FIBER ENHANCEMENT

The reinforcing elements (REINF263/264/265) are available in ANSYS190 to simulate the reinforcing fibers [1]. Reinforcing modeling in ANSYS190 has two options:

- Discrete modeling (SECTYPE, SECID, REINF, and DISC)—suitable for modeling reinforcing fibers with nonuniform materials or arbitrary orientations (Figure 9.1a).
- Smeared modeling (SECTYPE, SECIND, REINF, and SMEAR)—appropriate for modeling reinforcing fibers with identical materials and orientations (Figure 9.1b).

When the base elements are regular and the reinforcing location complies with the base elements, the standard method is used to create the reinforcing elements, including the following steps:

1. Create the base elements.
2. Build the reinforcing sections with respect to the base elements.
3. Select the base elements for reinforcement.
4. Generate the reinforcing elements using the EREINE command.

9.1.2 IVD MODEL WITH FIBER ENHANCEMENT

About 5 million Americans suffer from lower back pain [2]. The total cost of lower-back disabilities in the United States is approximately $50 billion annually. Three-quarters of lower back pain cases are associated with lumbar disc degeneration due to excessive physical activity. Thus, the mechanics of load transferred in the intervertebral disc (IVD) play an important role in understanding the patterns and mechanisms of back pain. Various finite element models have been built to gain a better understanding of the load distribution in the disc [3–7]. As mentioned previously, the annulus in the disc is enhanced by fibers that cross each other at an angle of about 150° (Figure 9.2). Implementation of collagen fibers in the model follows a number of approaches. Elliot and Setton built a linear, fiber-induced anisotropic model for

69

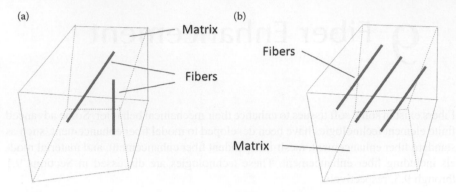

FIGURE 9.1 Schematic of the standard fiber enhancement. (a) Discrete modeling; (b) smeared modeling.

annulus fibrosus [8], while Wagner and Lotz developed a nonlinear elastic material model to present the annulus fibrosus [9]. In this section, we implemented the IVD model using fiber enhancement in ANSYS190.

9.1.2.1 Finite Element Model of IVD
9.1.2.1.1 Geometry and Mesh
The finite element model of half an IVD was built; assuming that it is symmetrical, the nucleus is in the center, enclosed by the annulus and covered by the end plate from the top and bottom (Figure 9.3). The annulus and end plate were meshed by SOLID185, and the nucleus was modeled by FULID80, as it was regarded as incompressible fluid.

9.1.2.1.2 Material Properties
The material properties of the annulus, nucleus, and end plate are listed in Table 9.1. The thermal expansion coefficient 0.001/°C was assigned to the nucleus for simulating the swelling of the nucleus.

A total of 15 layers of fibers were created in the annulus with the Young's modulus of each layer starting at 44 MPa, and an 8-MPa increase per layer added gradually

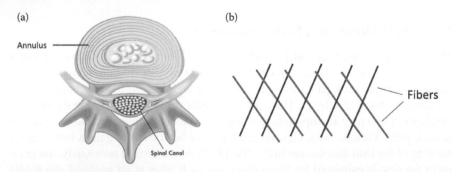

FIGURE 9.2 Fiber orientation in annulus. (a) IVD (Roberto Biasini,©123RF.com); (b) fiber orientation in annulus.

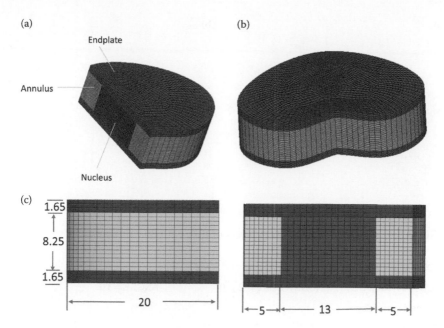

FIGURE 9.3 Finite element model of IVD. (a) Half a model; (b) full model; (c) dimensions of the IVD (all dimensions in millimeters).

from inside to outside [10]. Each layer of fibers was oriented at a 30° angle to the horizontal axis of the disc; this pattern alternated with a successive layer of fibers (Figure 9.4). The fiber cross section was selected as $100\,\mu m^2$ with a fiber gap of 10 μm. In addition, the fibers were defined as tension only using command SECC, 1:

```
MP, EX, 4, 156      ! material properties of fibers outside of annulus
MP, NUXY, 4, 0.1
MP, EX, 5, 148
MP, NUXY, 5, 0.1
......
MP, EX, 18, 44      ! material properties of fibers inside of annulus
MP, NUXY, 18, 0.1
! cross area AA =1 e-4mm^2; distances between fibers SS = 0.01 mm
```

TABLE 9.1

Material Properties of IVD

	Bone	Nucleus	Annulus
Young's modulus (MPa)	223.8	0.4	1.0
Poisson's ratio	0.4		0.48
Thermal expansion coefficient (1/°C)	0	0.001	0

(a)   (b)

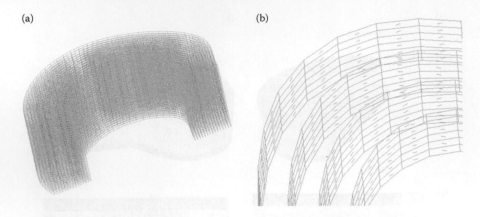

FIGURE 9.4   Fibers in the IVD; (a) 15 layers of fibers; (b) fiber orientation.

```
SECT, 1, REINF, SMEAR
SECD, 4, AA, SS, , ANGLE, ELEF, 2, 0          ! ANGLE=30°
! tension only
SECC, 1
SECT, 2, REINF, SMEAR
SECD, 5, AA, SS, , ANGLE1, ELEF, 2, 0         ! ANGLE1=150°
SECC, 1
......
```

9.1.2.1.3 Creating Fiber Elements

The annulus elements associated with each layer of the fibers were selected as base elements. One layer of fiber elements was created on the base elements using the command EREINF. The procedure was repeated for each of the 15 layers of fibers to create all the fiber elements:

```
! elements for fiber
*DO, J, 1, 15
SECN, J
ESEL, S, ELEM, ,1079 + (J − 1)*40, 1106 + (J − 1)*40
*DO, I, 1, 43
ESEL, A, ELEM, ,1079 + (J − 1)*40 + I*600, 1106 + I*600 + (J − 1)*40
*ENDDO
EREINF, ALL
ALLSEL
*ENDDO
ALLSEL
```

9.1.2.1.4 Loading and Boundary Conditions

The body temperature of the disc was increased 100°C to simulate swelling of the nucleus (a 30% volume increase), as other parts have a thermal expansion coefficient of zero. The bottom surface of the end plate was fixed in all degrees of freedom (DOFs).

9.1.2.2 Results

Thermal loading was applied to the nucleus to simulate the swelling of the nucleus. About 30% enlargement of the nucleus caused the end plate and the annulus close to the nucleus to become deformed (Figure 9.5). As the bone is much stiffer than the soft tissues, the major stresses occurred in the end plate (Figure 9.6). The major stresses of the annulus appeared in the area adjacent to the nucleus. The swelling of the nucleus also made most of the fibers in tension with the average axial stress around 1.85 MPa and the maximum stress at 3.33 MPa (Figure 9.7).

9.1.2.3 Discussion

A total of 15 layers of fiber were modeled with tension only in the IVD model, with various Young's moduli and alternative fiber directions. The deformation and stress distribution of the IVD in the swelling test looked reasonable, which confirmed the built model.

In the built model, the soft tissues were assumed to be solid only. Therefore, the obtained results such as deformation and stresses were the instantaneous responses. To study the longtime response of the IVD, the soft tissues should be modeled as porous media that is the topic of Section 11.3.

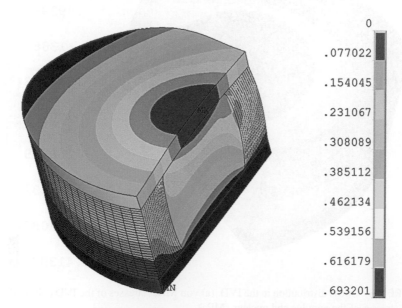

| 0 |
| .077022 |
| .154045 |
| .231067 |
| .308089 |
| .385112 |
| .462134 |
| .539156 |
| .616179 |
| .693201 |

FIGURE 9.5 Deformation of the IVD after 30% swelling of the nucleus (mm).

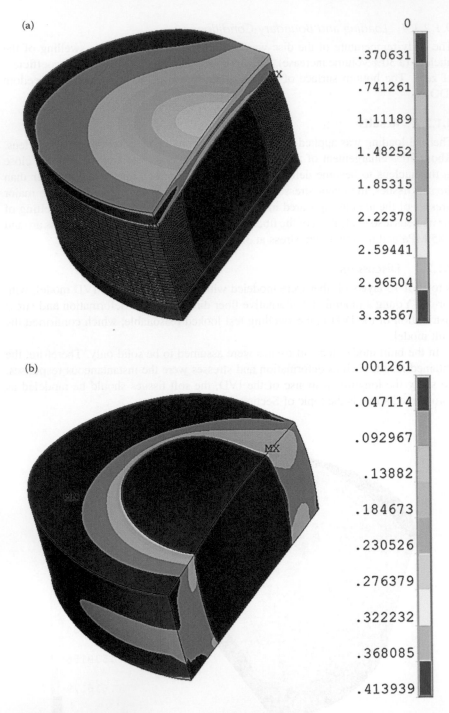

FIGURE 9.6 Stress distribution in the IVD. (a) von Mises stresses of the IVD (MPa); (b) von Mises stresses of the annulus and nucleus (MPa).

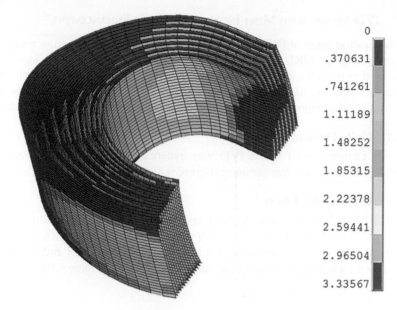

FIGURE 9.7   Fiber axial stresses (MPa).

9.1.2.4   Summary

Fiber enhancement in the IVD was simulated in ANSYS190, with material properties and fiber orientation close to the reality. The model can be used to study the load distribution in the disc.

9.2   MESH-INDEPENDENT FIBER ENHANCEMENT

9.2.1   INTRODUCTION OF MESH-INDEPENDENT FIBER ENHANCEMENT

When the base elements are arbitrary and the reinforcing cannot be performed following distinct patterns, the mesh-independent method is an alternative [1]. The basic idea for the mesh-independent method is to create a geometry for the reinforcing part and mesh it with MESH200 to generate MESH200 elements that are independent of the base elements. Then the reinforcing elements are created within MESH200 elements and base elements. The mesh-independent method involves the following steps:

1. Create the base elements.
2. Build the reinforcing geometry and mesh it with MESH200.
3. Select the base elements and the MESH200 elements.
4. Generate the reinforcing elements using the EREINF command.

9.2.2 IVD Model with Mesh-Independent Fiber Enhancement

In the biomedical study of IVD, the geometry of IVD always comes from a magnetic resonance imaging (MRI) or computed tomography (CT) scan. Free meshing is widely used to mesh the annulus because of its irregular geometry. In this situation, the fiber enhancement method described in Section 9.1.2 does not work for building fibers within the annulus. One alternative is mesh-independent fiber enhancement.

9.2.2.1 Finite Element Model

The finite element model of the IVD was generated with the same geometry as Section 9.1.2, but it used free meshing (Figure 9.8a).

9.2.2.2 Creating the Fibers

A total of 15 surfaces represent the fiber layers (Figure 9.8b). These surfaces were meshed with MESH200 and assigned with the material identity of each layer. Then all elements were selected, including the base element SOLID285 and MESH200, to generate the fiber elements using the command EREINF (Figure 9.9).

```
ET, 5, 200, 6
TYPE, 5
! cross area AA =1 e-4mm^2; distances between fibers SS = 0.01mm
! create fibers with angle 30°
*DO, J, 1, 15, 2
SECT, J, REINF, SMEAR
SECD, J + 3, AA, SS, ,ANGLE, MESH
! tension only
SECC, 1
ASEL, S, AREA, , 9 + J
ESIZE, 1
SECN, J
AMESH, 9 + J
ALLSEL
*ENDDO
! create fibers with angle 150°
*DO, J, 2, 14, 2
SECT, J, REINF, SMEAR
SECD, J+3, AA, SS, ,ANGLE1, MESH
! tension only
SECC, 1
ASEL, S, AREA, ,9 + J
ESIZE, 1
```

(a)

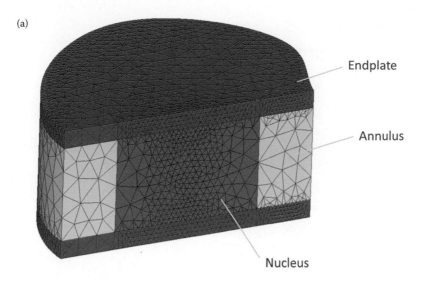

Endplate

Annulus

Nucleus

(b)

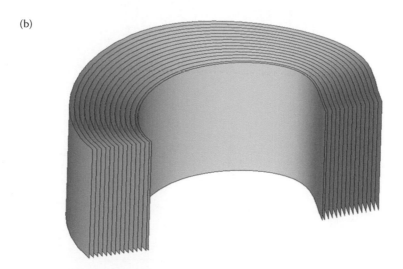

FIGURE 9.8   Finite element model of the IVD with free meshing. (a) IVD model; (b) fiber planes.

```
SECN, J
AMESH, 9 + J
ALLSEL
*ENDDO
ESEL, ALL
CSYS, 0
EREINF
```

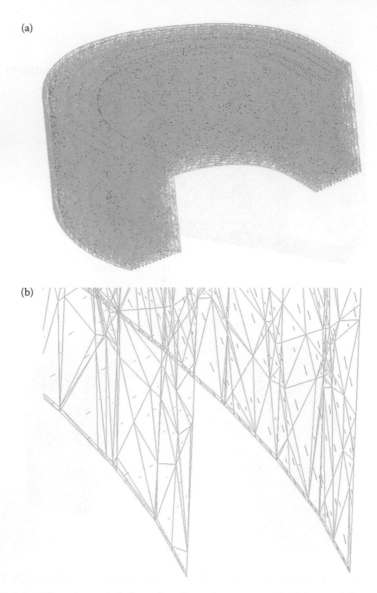

FIGURE 9.9 Fibers by mesh-independent fiber enhancement. (a) 15 layers of fibers; (b) fiber orientation.

9.2.2.3 Results

The problem was solved with the same loading and boundary conditions as in Section 9.1.2. Figures 9.10 through 9.12 illustrate the deformation and von Mises stresses of the IVD, as well as the axial stress distribution of the fibers. These results are very close to those in Section 9.1.2, which validates the mesh-independent method.

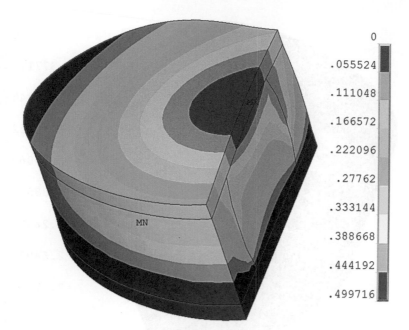

0
.055524
.111048
.166572
.222096
.27762
.333144
.388668
.444192
.499716

FIGURE 9.10 Deformation of the IVD (mm).

9.2.2.4 Summary

The fibers were created within irregular annulus elements using MESH200. The obtained results match those of the regular meshing given in Section 9.1.2, which demonstrates the capability of the mesh-independent method. Because most practical problems are meshed by free meshing, this new technology has a much wider application than the mesh-dependent method.

9.3 MATERIAL MODELS INCLUDING FIBER ENHANCEMENT

9.3.1 Anisotropic Material Model with Fiber Enhancement

The anisotropic material model was developed to simulate elastomers with reinforcements, such as arteries in biomedical engineering [1]. It has two options. The first, the polynomial-function-based strain energy potential, has the strain energy density function defined by

$$W = W_v(J) + W_d(\bar{C}, \boldsymbol{A} \otimes \boldsymbol{A}, \boldsymbol{B} \otimes \boldsymbol{B}) \tag{9.1}$$

$$W_v(J) = \frac{1}{d}(J-1)^2 \tag{9.2}$$

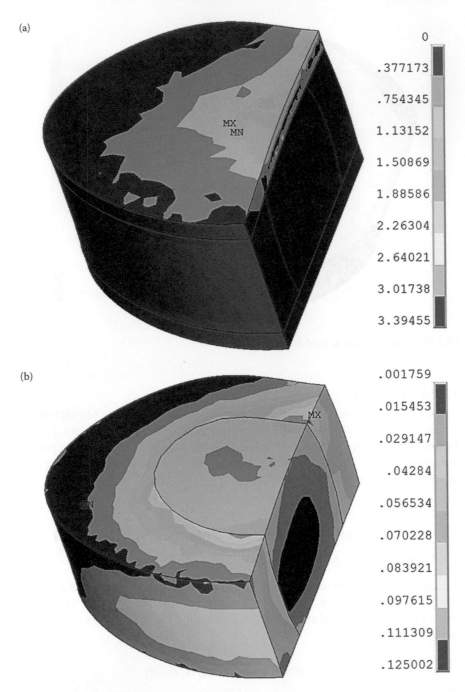

FIGURE 9.11   von Mises stresses of the IVD. (a) von Mises stresses of the IVD (MPa); (b) von Mises stresses of the annulus and nucleus (MPa).

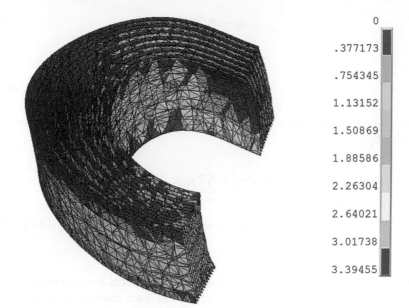

FIGURE 9.12 Fiber axial stresses (MPa).

$$W_d(\bar{\boldsymbol{C}}, \boldsymbol{A} \otimes \boldsymbol{A}, \boldsymbol{B} \otimes \boldsymbol{B}) = \sum_{i=1}^{3} a_i(I_1 - 3)^i + \sum_{j=1}^{3} b_j(I_2 - 3)^j$$

$$+ \sum_{k=2}^{6} c_k(I_4 - 1)^k + \sum_{l=2}^{6} d_l(I_5 - 1)^l$$

$$+ \sum_{m=2}^{6} e_m(I_6 - 1)^m + \sum_{n=2}^{6} f_n(I_7 - 1)^n + \sum_{o=2}^{6} g_o(I_8 - \varsigma)^o,$$

$$(9.3)$$

where

$$\varsigma = (\boldsymbol{A} \cdot \boldsymbol{B})^2 \tag{9.4}$$

$$I_1 = \text{tr } \boldsymbol{C} \tag{9.5}$$

$$I_2 = \frac{1}{2}\text{tr } \boldsymbol{C}^2 \tag{9.6}$$

$$I_4 = \boldsymbol{A} \cdot \boldsymbol{CA} \tag{9.7}$$

$$I_5 = \boldsymbol{A} \cdot \boldsymbol{C}^2\boldsymbol{A} \tag{9.8}$$

$$I_6 = \mathbf{B} \cdot \mathbf{CB} \tag{9.9}$$

$$I_7 = \mathbf{B} \cdot \mathbf{C}^2\mathbf{B} \tag{9.10}$$

$$I_8 = (\mathbf{A} \cdot \mathbf{B})\mathbf{A} \cdot \mathbf{CB} \tag{9.11}$$

The second option, the exponential function–based strain energy potential, is expressed by

$$W_d(\bar{C}, \mathbf{A} \otimes \mathbf{A}, \mathbf{B} \otimes \mathbf{B}) = \sum_{i=1}^{3} a_i (I_1 - 3)^i + \sum_{j=1}^{3} b_j (I_2 - 3)^j$$
$$+ \frac{c_1}{2c_2} \{ \exp[c_2 (I_4 - 1)^2] - 1 \} + \frac{e_1}{2e_2} \{ \exp[e_2 (I_6 - 1)^2] - 1 \}. \tag{9.12}$$

Eqs. (9.3) and (9.12) indicate that the anisotropic material is composed of a base material $\mathbf{C}$ enhanced by two fibers $\mathbf{A}$ and $\mathbf{B}$. The base material $\mathbf{C}$ is almost the same as that with the two-parameter Yeoh model.

The following APDL commands are used to define the anisotropic material model with the Polynomial option in ANSYS190:

```
TB, AHYPER, , , POLY
TBDATA, 1, A1, A2, A3, B1, B2, B3...          ! total 31 parameters
TB, AHYPER, , ,AVEC
TBDATA, 1, A1, A2, A3
TB, AHYPER, , ,BVEC
TBDATA, 1, B1, B2, B3
TB, AHYPER, , ,PVOL
TBDATA, 1, D
```

The anisotropic material model with the exponential option is defined by

```
TB, AHYPER, , ,EXP
TBDATA, 1, A1, A2, A3, B1, B2, B3
TBDATA, 7, C1, C2, C3 ! total 10 parameters
TB, AHYPER, , ,AVEC
TBDATA, 1, A1, A2, A3
TB, AHYPER, , ,BVEC
TBDATA, 1, B1, B2, B3
TB, AHYPER, , ,PVOL
TBDATA, 1, D
```

To illustrate the differences among the terms I_4, I_6, I_5, and I_7, one uniaxial tension test with material **A** and material **B** was conducted. Both materials have the same neo-Hookean model as the base material and are enhanced by fibers in the z-direction. The only difference between the two materials is that the terms I_6 and I_7 are used for fiber effects in materials **A** and **B**, respectively. Figure 9.13 shows the mechanical responses of the two materials with different terms under the same displacement loading in the z-direction. The results show that material **B**, using term I_7, has bigger stresses than material **A** using term I_6, although both have the same loading conditions.

```
! Anisotropic Hyperelastic Material A using term I6
TB, AHYPER, 1, 1, 31, POLY
TBDATA, 1, 3
TBDATA, 17, 2                    ! using term I6
TB, AHYPER, 1, 1, 3, AVEC
TBDATA, 1, 1, 0, 0
TB, AHYPER, 1, 1, 3, BVEC
TBDATA, 1, 0, 0, 1

! Anisotropic Hyperelastic Material B using term I7
TB, AHYPER, 1, 1, 31, POLY
TBDATA, 1, 3
TBDATA, 22, 2                    ! using term I7
TB, AHYPER, 1, 1, 3, AVEC
TBDATA, 1, 1, 0, 0
```

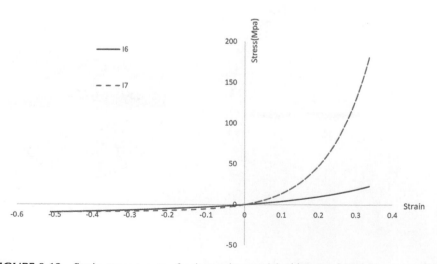

FIGURE 9.13 Strain-stress curves of anisotropic material with terms I_6 and I_7.

```
TB, AHYPER, 1, 1, 3, BVEC
TBDATA, 1, 0, 0, 1
```

Figure 9.13 also illustrates the various responses in the events of tension and compression, which indicate that the fibers in the anisotropic model are tension-only.

9.3.2 Simulation of Anterior Cruciate Ligament (ACL)

The anterior cruciate ligament (ACL) plays a crucial role in the knee. It prevents the anterior tibial translation and bears about 90% of the anteriorly directed load applied to the tibia between the 30 degrees and 90 degrees of flexion. The ACL is the most injured ligament of the human body (Figure 9.14) [11,12], considering that it has been estimated that there are almost 100,000 cases in the United States every year [13]. A sound knowledge of stress and strain distributions within the ACL is indispensable for understanding the causes of ACL injuries, the consequences, and the ways to prevent such injuries from occurring. Finite element models can provide profound insights into the mechanical characteristics of the ACL, which are very difficult or even impossible to assess by experimentation. Many full three-dimensional (3D) finite element models of the ACL have been developed [14–19]. In this section, a 3D ACL with anisotropic material model was built in ANSYS190.

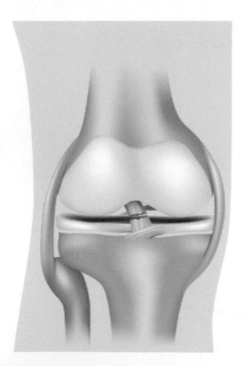

FIGURE 9.14   ACL rupture (alila © 123RF.com).

9.3.2.1 Finite Element Model

9.3.2.1.1 Geometry and Mesh

The finite element analysis was performed on the knee model from Open Knee [20] (under the terms of the Creative Commons Attribution 3.0 License). This study focused on 3D mechanical behavior of the ACL. Thus, the finite element model in this study comprises only three parts: the femur, tibia, and ACL (Figure 9.15).

The femur and tibia were meshed with SHELL181, and the ACL was modeled with SOLID185.

9.3.2.1.2 Material Model

The ACL material was modeled by the anisotropic material model with exponential anisotropic strain energy potential [21]:

```
TB, AHYPER, 1, ,10, EXP
TBDATA, 1, 1.5
TBDATA, 7, 4.39056, 12.1093
TB, AHYPER, 1, , ,AVEC
TBDATA,1, 1, 0, 0                    ! fiber enhancement in the x direction
of local ESYS
TB, AHYPER, 1, , ,PVOL
TBDATA, 1, 1e-3
```

The bone was assumed to be isotropic, with a Young's modulus of 1,000 MPa and Poisson's ratio of 0.3.

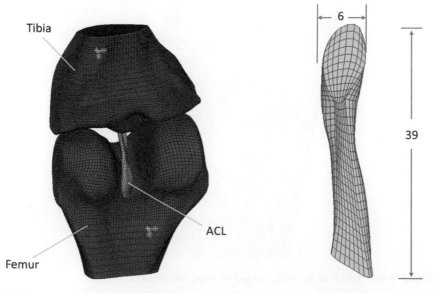

FIGURE 9.15 Finite element model of ACL (all dimensions in millimeters).

9.3.2.1.3 Element Coordinate System

Like the simulations of the bone with anisotropic materials described in Chapter 4, element coordinate systems are required for all elements of the ACL, as the ACL is anisotropic (Figure 9.16). Thus, the local coordinate system 12 was defined and assigned to all ACL elements:

```
CS, 12, 0, 95123, 95128, 95078, 1, 1,

ESEL, S, TYPE, ,1

EMODIF, ALL, ESYS, 12

ALLSEL

CSYS,0
```

9.3.2.1.4 Boundary Conditions

A pilot node was defined to couple with the femur and control its motion (Figure 9.17a). The displacement loading that the femur moves 4 mm in the x-direction was applied to the pilot node 100001.

Similarly, coupling between a pilot node and the tibia was set up, and all the DOFs of the pilot node were constrained to fix the tibia (Figure 9.17b).

Connection of the ACL with the bones was implemented in ANSYS using the CP command (Figure 9.18).

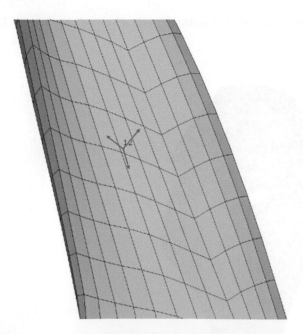

FIGURE 9.16 ESYS of the ACL, defined by local coordinate system 12.

(a)   (b)

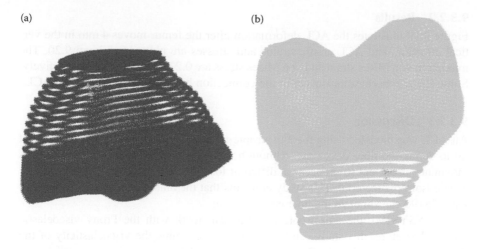

FIGURE 9.17 MPC constraints on the bone (a) MPC constraints on tibia; (b) MPC constraints on femur.

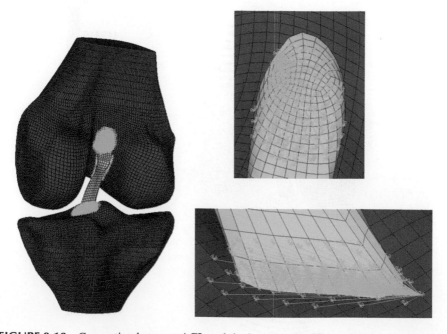

FIGURE 9.18 Connection between ACL and the bone via the CP command.

9.3.2.2 Results

Figure 9.19 illustrates the ACL deformation after the femur moves 4 mm in the vertical direction, and ACL elastic strains and stresses are plotted in Figure 9.20. The maximum von Mises strain and von Mises stress are 0.23 and 3.32 MPa, respectively. Some stress concentrations occur at the connection between the bone and the ACL.

9.3.2.3 Discussion

An ACL was modeled using the anisotropic material model. A local system was set up to define the fiber directions. Although the ACL was assumed to be elastic, its maximum stress and strain occur at different locations, which is not the same as the elastic isotropic material. This study confirms that the anisotropic material model in ANSYS190 can simulate the fiber enhancement.

In ANSYS190, the anisotropic material can work with the Prony viscoelastic model. Therefore, this study can be extended to examine the viscoelasticity of the ACL after incorporating the Prony viscoelastic model.

9.3.2.4 Summary

An ACL was modeled by the anisotropic material model with exponential anisotropic strain energy potential. The developed finite element model can be used to study the viscoelasticity of the ACL by combining the Prony viscoelastic model with the anisotropic material model.

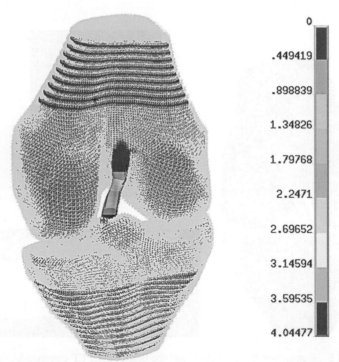

0
.449419
.898839
1.34826
1.79768
2.2471
2.69652
3.14594
3.59535
4.04477

FIGURE 9.19 Deformation of ACL (mm).

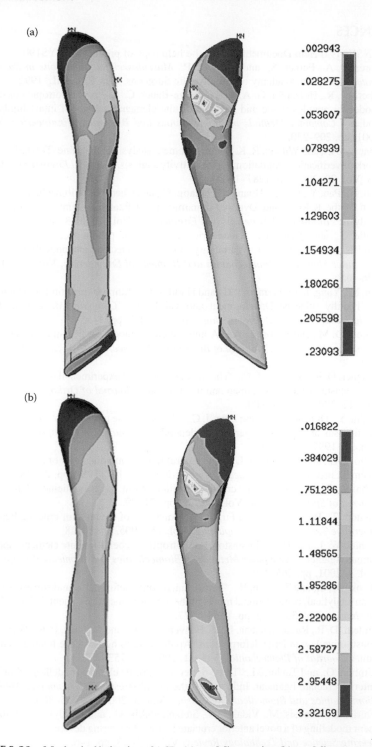

FIGURE 9.20 Mechanical behavior of ACL. (a) von Mises strains; (b) von Mises stresses (MPa).

REFERENCES

1. ANSYS19.0 Help Documentation in the help page of product ANSYS190.
2. Paremer, A., Fumer, S., and Rice, D. P., *Musculoskeletal Conditions in the United States*. American Academy of Orthopaedic Surgeons, Park Ridge, IL, 1992.
3. Eberline, R., Holzapfel, G. A., and Schulze-Bauer, C. A., "An anisotropic constitutive model for annulus tissue and enhanced finite element analyses of intact lumbar disc bodies." *Computer Methods in Biomechanics and Biomedical Engineering*, Vol. 4, 2001, pp. 209–230.
4. Jones, A. C., and Wilcox, R. K., "Finite element analysis of the spine: Towards a framework of verification, validation, and sensitivity analysis." *Medical Engineering & Physics*, Vol. 30, 2008, pp. 1287–1304.
5. Lin, H., Pan, Y., Liu, C., Huang, L., Huang, C., and Chen, C., "Biomechanical comparison of the K-ROD and Dynesys dynamic spinal fixator systems—a finite element analysis." *Bio-Medical Materials and Engineering*, Vol. 23, 2013, pp. 495–505.
6. Little, J., and Adam, C., "Geometric sensitivity of patient-specific finite element models of the spine to variability in user selected anatomical landmarks." *Computer Methods in Biomechanics and Biomedical Engineering*, Vol. 18, 2013, pp. 676–688.
7. Xu, M., Yang, J., Lieberman, J. H., and Haddas, R., "Lumbar spine finite element model for healthy subjects: Development and validation." *Computer Methods in Biomechanics and Biomedical Engineering*, Vol. 20, 2017, pp. 1–15.
8. Elliott, D. M., and Setton, L. A., "A linear material model for fiber-induced anisotropy of the annulus fibrosus." *Journal of Biomedical Engineering*, Vol. 122, 2000, pp. 173–179.
9. Wagner, D. R., and Lotz, J. C., "Theoretical model and experimental results for the nonlinear elastic behavior of human annulus fibrosus." *Journal of Orthopaedic Research*, Vol. 22, 2006, pp. 901–909.
10. Fujita, Y., Duncan, N. A., and Lotz, J. C., "Radial tensile properties of the lumbar annulus fibrosus are site and degeneration dependent." *Journal of Orthopaedic Research*, Vol. 15, 1997, pp. 814–819.
11. Fetto, J. E., and Marshall, J. L., "The natural history and diagnosis of anterior cruciate ligament insufficiency." *Clinical Orthopaedics*, Vol. 147, 1980, pp. 29–38.
12. Johnson, R. J., "The anterior cruciate: A dilemma in sports medicine." *International Journal of Sports Medicine*, Vol. 3, 1982, pp. 71–79.
13. Zantop, T., Petersen, W., and Fu, F. H., "Anatomy of the anterior cruciate ligament." *Operative Techniques in Orthopaedics*, Vol. 15, 2005, pp. 20–28.
14. Daniel, W. J. T., "Three-dimensional orthotropic viscoelastic finite element model of a human ligament." *Computer Methods in Biomechanics and Biomedical Engineering*, Vol. 4, 2001, pp. 265–279.
15. Hirokawa, S., and Tsuruno, R., "Three-dimensional deformation and stress distribution in an analytical/computational model of the anterior cruciate ligament." *Journal of Biomechanics*, Vol. 33, 2000, pp. 1069–1077.
16. Pioletti, D. P., Rakotomanana, L. R., Benvenuti, J. F., and Leyvraz, P. F., "Viscoelastic constitutive law in large deformations: Application to human knee ligaments and tendons." *Journal of Biomechanics*, Vol. 31, 1998, pp. 753–757.
17. Limbert, G., and Taylor, M., "Three-dimensional finite element modelling of the human anterior cruciate ligament. Influence of the initial stress field." *Computer Methods in Biomechanics and Biomedical Engineering*, Vol. 3, 2001, pp. 355–360.
18. Vairis, A., Petousis, M., Vidakis, N., Stefanoudakis, G., and Kandyla, B., "Finite element modelling of a novel anterior cruciate ligament repairing device." *Journal of Engineering Science and Technology Review*, Vol. 6, 2013, pp. 1–6.

19. Parekh, J. N., *Using Finite Element Methods to Study Anterior Cruciate Ligament Injuries: Understanding the Role of ACL Modulus and Tibial Surface Geometry on ACL Loading*, University of Michigan, 2013.

20. Open Knee(s): Virtual Biomechanical Representations of the Knee Joint. Website: simtk.org/projects/openknee.

21. Peña, E., Calvo, B., Martínez, M. A., and Doblaré, M., "An anisotropic visco-hyperelastic model for ligaments at finite strains. Formulation and computational aspects." *International Journal of Solids and Structures*, Vol. 44, 2007, pp. 760–778.

10 USERMAT for Simulation of Soft Tissues

The materials in biology always exhibit as strongly nonlinear. However, the strain energy density functions of soft tissues obtained by experiments sometimes are much different from those defined in ANSYS. Therefore, it is necessary for users to write their own hyperelastic law and implement it in ANSYS.

10.1 INTRODUCTION OF SUBROUTINE USERHYPER

The subroutine UserHyper should follow the defined format [1]:

```
subroutine UserHyper(
        &              prophy, incomp, nprophy, invar,
        &              potential, pInvDer)
```

In these input and output parameters, prophy and nprophy are inputs of material parameters; invar refers to the input of invariants I_1, I_2, and J; and potential W and pInvDer are outputs.

pInvDer is a derivative of the strain-energy potential with respect to I_1, I_2, and J, including nine terms $(\partial W/\partial I_1)$, $(\partial W/\partial I_2)$, $(\partial^2 W/\partial I_1 \partial I_1)$, $(\partial^2 W/\partial I_1 \partial I_2)$, $(\partial^2 W/\partial I_2 \partial I_2)$, $(\partial^2 W/\partial I_1 \partial J)$, $(\partial^2 W/\partial I_2 \partial J)$, $(\partial W/\partial J)$, and $(\partial^2 W/\partial J \partial J)$, numbered consecutively from 1 to 9.

10.2 SIMULATION OF AAA USING USERHYPER

10.2.1 Using Subroutine UserHyper to Simulate Soft Tissues of the Artery

The strain energy density function of soft tissues was assumed to have the exponential form [2]

$$W(I_1) = \frac{\alpha}{2\gamma}(e^{\gamma(I_1-3)} - 1) + \frac{1}{d}\left(\frac{J^2 - 1}{2} - \ln J\right) \tag{10.1}$$

Calculate the first and second derivatives of the strain-energy potential with respect to these three invariants as follows:

$$\frac{\partial W}{\partial I_1} = \frac{\alpha}{2}e^{\gamma(I_1-3)} \tag{10.2}$$

$$\frac{\partial W}{\partial J} = \frac{1}{d}\left(J - \frac{1}{J}\right) \tag{10.3}$$

93

$$\frac{\partial^2 W}{\partial I_1 \partial I_1} = \frac{\alpha\gamma}{2} e^{\gamma(I_1-3)} \tag{10.4}$$

$$\frac{\partial^2 W}{\partial J \partial J} = \frac{1}{d}\left(1 + \frac{1}{J^2}\right) \tag{10.5}$$

The other derivatives are zero. Thus, these equations was implemented in the UserHyper subroutine:

```
Subroutine UserHyper(
     &      prophy, incomp, nprophy, invar,
     &      potential, pInvDer)
#include "impcom.inc"
      DOUBLE PRECISION ZERO, ONE, TWO, THREE, HALF, TOLER
      PARAMETER   (ZERO = 0.d00,
     &         ONE = 1.0d0,
     &         HALF = 0.5d0,
     &         TWO = 2.d0,
     &         THREE = 3.d0,
     &         TOLER = 1.0d-12)
      INTEGER      nprophy
      DOUBLE PRECISION prophy(*), invar(*),
     &         potential, pInvDer(*)
      LOGICAL      incomp
      DOUBLE PRECISION i1, jj, a1, g1, oD1, j1
c

      i1 = invar(1)
      jj = invar(3)
      a1 = prophy(1)
      g1 = prophy(2)
      oD1 = prophy(3)
      potential = ZERO
      pInvDer(1) = ZERO
      pInvDer(3) = ZERO

      potential = a1/TWO/g1 * (exp(g1 * (i1 - THREE)) - ONE)
      pInvDer(1) = a1/TWO * exp(g1 * (i1 - THREE))
```

```
pInvDer(3) = g1 * pInvDer(1)
j1 = ONE / jj
pInvDer(8) = ZERO
pInvDer(9) = ZERO
IF(oD1 .gt. TOLER) THEN
oD1 = ONE / oD1
incomp = .FALSE.

potential = potential + oD1 * ((jj*jj - ONE)*HALF - log(jj))
pInvDer(8) = oD1 * (jj - j1)
pInvDer(9) = oD1 * (ONE + j1 * j1)
END IF
```

c

```
RETURN
END
```

10.2.2 VALIDATION

The Taylor series for the exponential function e^x at $x = 0$ is

$$e^x = 1 + x + \frac{x^2}{2} + \frac{x^3}{6} + \cdots \tag{10.6}$$

Thus, by substituting Eq. (10.6) into Eq. (10.1), the strain energy density function becomes

$$W(I_1) = \frac{\alpha}{2}(I_1 - 3) + \frac{\alpha\gamma}{4}(I_1 - 3)^2 + \frac{\alpha\gamma^2}{12}(I_1 - 3)^3 + \frac{1}{d}\left(\frac{J^2 - 1}{2} - \ln J\right) \tag{10.7}$$

This matches the strain energy density function of the nine-parameter Mooney-Rivlin [Eq. (7.4)]. Therefore, Eq. (10.1) can be approximated by the nine-parameter Mooney-Rivlin.

A uniaxial tension test was conducted with equivalent material properties defined by tb, hyper, , , ,user and tb, hyper, , , ,mooney, respectively:

```
! define material by tb, hyper, , , ,user
TB, HYPER, 1, ,3, USER
TBDATA, 1, 0.12, 0.6, 0

! define material by tb, hyper, , , ,Mooney
TB, HYPER, 1, ,9, MOONEY
TBDATA, 1, 0.06, 0.0, 0.03*0.6, 0.0, 0.0, 0.01*0.6*0.6
```

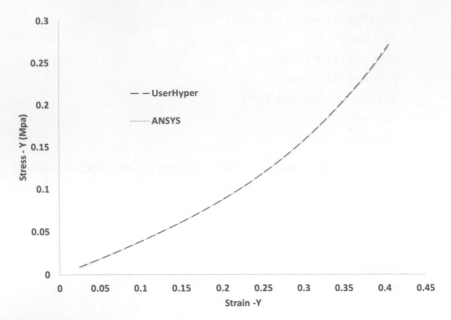

FIGURE 10.1 Comparison of UserHyper with the ANSYS nine-parameter Mooney-Rivlin model.

The stresses obtained by `tb,hyper,,,,user` and `tb,hyper,,,,mooney` match very well (Figure 10.1), which validates the developed UserHyper.

10.2.3 STUDY THE AAA USING USERHYPER

In Section 7.2 of Chapter 7, the Ogden material model was utilized to simulate the wall stresses of abdominal aortic aneurysm (AAA) (Figure 10.2). In this section, AAA was simulated by the following material model defined by UserHyper [3]:

```
! define material by tb, hyper, , , ,user
TB, HYPER, 1, ,3, USER
TBDATA, 1, 310000E-6, 1.87, 0
```

After solving this, the obtained deformation and von Mises stresses are plotted in Figures 10.3 and 10.4. The deformation is asymmetrical, with a maximum value of 5.36 mm. The major stresses occur at the transition of the aorta and aneurysm with a peak value of 0.38 MPa.

10.2.4 DISCUSSION

The soft tissues of the artery have the exponential form of the strain energy density function that has no corresponding hyperelastic model in ANSYS. Therefore, UserHyper was developed to simulate it.

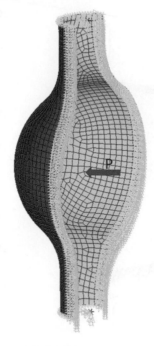

FIGURE 10.2   Finite element model of AAA.

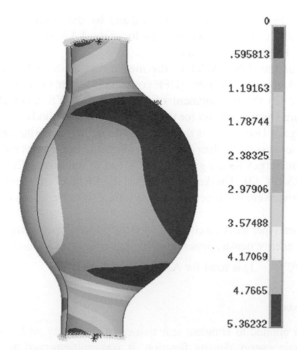

FIGURE 10.3   Deformation of the AAA wall (mm).

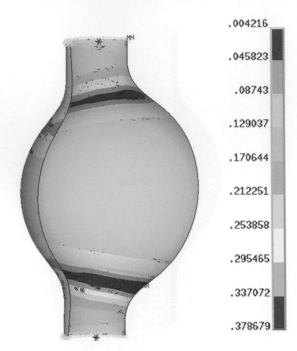

FIGURE 10.4   von Mises stresses of the AAA wall (MPa).

The new developed UserHyper was validated by the close form of the nine-parameter Mooney-Rivlin and then applied to study AAA. This demonstrates the capability of the UserHyper of ANSYS.

Three methods link USERMAT to the mechanical ANSYS Parametric Design Language (APDL) program: (1) the /UPF command; (2) the creation of a dynamic-link library; and (3) the development of a custom ANSYS executable. Refer to ANSYS19.0 Help Documentation for the details of three methods.

One advantage of UserHyper is that it is easy to implement. The standard USER-MAT requires the stresses and their derivatives to output. Sometimes it is very difficult to compute them. Sergey took a great deal of effort to implement Eq. (9.1) with the standard USERMAT [3]. Unlike the standard USERMAT, however, UserHyper just needs to provide energy potential and its nine derivatives, which are very easy to implement.

Although UserHyper works well with Eq. (10.1), UserHyper cannot work with all forms of strain energy density functions. For example, the simulation fails in case $W(I_1) = \dfrac{\alpha}{2\gamma}(e^{\gamma(I_1-3)^2} - 1)$ is used for AAA material.

10.2.5 Summary

UserHyper was applied to simulate soft tissues of the artery with the exponential form of the strain energy density function. It was implemented and validated in ANSYS.

REFERENCES

1. ANSYS19.0 Help Documentation in the help page of product ANSYS190.
2. Wulandana, R., and Robertson, A. M., "An inelastic multi-mechanism constitutive equation for cerebral arterial tissue." *Biomechanics and Modeling in Mechanobiology,* 4, 2005, 235–248.
3. Sidorov, S. *Finite Element Modelling of Human Artery Tissue with a Nonlinear Multi-Mechanism Inelastic Material.* University of Pittsburgh, 2007.

11 Modeling Soft Tissues as Porous Media

As mentioned in Chapter 6, soft tissues are biphasic, containing 30%–70% of water. The internal friction between the fluid and solid phases greatly influences the mechanical behavior of soft tissues. Thus, coupled pore-pressure thermal (CPT) elements that model porous media in ANSYS can be applied to study soft tissues, and this topic is discussed in this chapter.

11.1 CPT ELEMENTS

Porous media are treated as multiphase materials following an extended version of Biot's consolidation theory [1]. Darcy's Law described the flow as

$$\mathbf{q} = -k\nabla p, \tag{11.1}$$

where $\mathbf{q}$ is flux, k is permeability, p is pore pressure, and ∇ is the gradient operator.

The solid phase is governed by the following equation:

$$\nabla \cdot (\boldsymbol{\sigma}'' - \alpha p\mathbf{I}) + f = 0, \tag{11.2}$$

where f is body force, α is the Biot coefficient, $\boldsymbol{\sigma}''$ is Biot effective stress, and $\mathbf{I}$ is the second-order identity tensor.

In the case where there is no flow source or temperature change, the mass balance equation for the fluid becomes

$$\nabla\mathbf{q} + \alpha\dot{\varepsilon}_v + \frac{\dot{p}}{Q^*} = 0, \tag{11.3}$$

where $\dot{\varepsilon}_v$ is volumetric strain and Q^* is the compressibility parameter.

The unknowns in Eqs. (11.1) through (11.3) are u (displacement of the solid phase), w (fluid displacement), and p (pore pressure). Thus, these equations can be solved directly. However, to make the computation more efficient, these three equations can be condensed into two equations without the unknown w, which is used in ANSYS to simulate the porous media. Therefore, CPT element types were developed in ANSYS, including CPT212 (two-dimensional (2D) 4-node); CPT213 (2D 8-node); CPT215 (three-dimensional (3D) 8-node); CPT216 (3D 20-node); and CPT217 (3D 10-node). The porous material properties are defined by the following commands:

```
TB, PM, , , , PERM

TBDATA, 1, 1e-8        ! define solid permeability k

TB, PM, , , , BIOT     ! define Biot coefficient

TBDATA, 1, 0.5
```

For consolidation problems, the initial time step for dynamical analysis can be defined as follows [2]

$$\Delta t \geq \frac{\gamma_w}{6Ek}(\Delta h)^2, \tag{11.4}$$

where E is the Young's modus of a solid skeleton, γ_w is the specific weight of fluid, and Δh is the element size.

11.2 STUDY OF HEAD IMPACT

Head injury (HI) incidence occurs in approximately 180–250 individuals per 100,000 people in the United States each year. Over half a million people are hospitalized because of HIs. The direct medical costs for treatment are estimated at about $50 billion per year [3]. Traffic accidents have the highest mortality rate of HI. Protection devices such as safety belts, air bags, and safety helmets have played a significant role in decreasing HI mortality, and yet improvements can always be made. Understanding the mechanisms of HI is the first step to improving the design of protective devices.

The finite element method has been applied to study HI as early as the 1970s [4]. Among numerous numerical simulations, three studies [5–7] are considered as milestones in finite element modeling of the human head. However, due to the extremely complex geometry and material properties of the human head, finite element modeling of the human head remains challenging. In this section, head impact was simulated in ANSYS, in which the soft tissues were considered as porous media.

11.2.1 Finite Element Model of the Head

11.2.1.1 Geometry and Mesh
The head was simplified to a 2D plane strain model comprised of the brain, the skull, and cerebrospinal fluid (CSF). The main part of the head contains the brain in the center, surrounded by CSF. The outer edge of the head is the skull, a layer of hard shell (Figure 11.1).

The brain and CSF were modeled as soft tissues meshed by CPT212, and the skull was regarded as solid and meshed by PLANE182.

11.2.1.2 Material Properties
The material of the skull was assumed to be elastic, isotropic, and homogeneous. Both the brain and CSF were modeled as porous, but the permeability of the CSF was set extremely high to simulate its free flow. The material models of the brain, CSF, and skull are listed in Table 11.1 [8].

11.2.1.3 Loading and Boundary Conditions
To simulate the impact, an 8,000-N force was applied on node 82 in 1e–3 s and unloaded in 1e–3 s (Figure 11.2). The mechanical behavior of the head

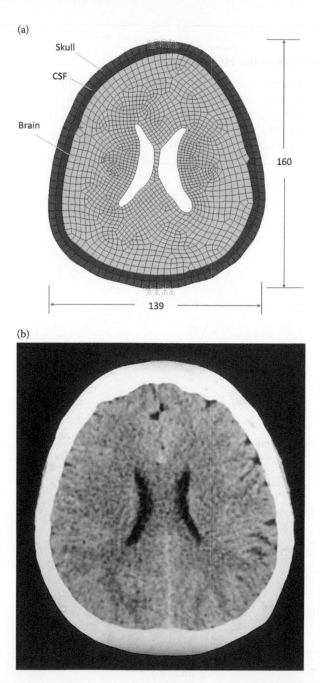

FIGURE 11.1 Finite element model of head impact. (a) finite element model of a head (all dimensions in millimeters); (b) CT of a head. (somkku9kanokwan © 123RF.com.)

TABLE 11.1

Material Properties of the Head Components

	Young's Modulus	Poisson's Ratio	Permeability $(m^4/N/s)$	Density (kg/m^3)
Skull	6.5e9	0.22	N/A	1,412
Brain	66,700	0.48	4.8e–8	1,040
CSF	66,700	0.499	1	1,040

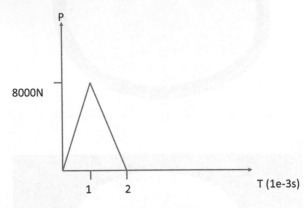

FIGURE 11.2   Loading history.

within 1e–1 s was examined. Therefore, three loading steps were defined in the solution setting:

```
/SOLU
ANTYPE, SOIL
KBC, 0                      ! ramp load
OUTRES, ALL, ALL
NLGEOM, ON
TIME, 1E-3
NSUBST, 100, 1000, 20
SOLV
TIME, 2E-3
F, 82, FX, 0               ! unload
ALLSEL
SOLV
TIME, 100E-3               ! no loading
SOLV
```

No rigid-body motion is allowed in the model. Thus, four nodes at the top and four nodes at the bottom were fixed in both the x- and y-directions to constrain the skull.

11.2.2 Results

The deformations at the end of each loading step are plotted in Figure 11.3. The main deformation occurs at the brain tissues because the skull is much stiffer than the brain tissues. Figure 11.4 illustrates the pore pressure of the CSF, in which the values are almost uniform. This is consistent with the CSF being regarded as fluid. The pore pressures of the brain tissues decrease with time, although their distribution greatly varies at each time step (Figure 11.5), which was confirmed by the time history of the pore pressure at node 582 (Figure 11.6). Both the pore pressures and the displacements at the x-direction decay with time (Figure 11.7).

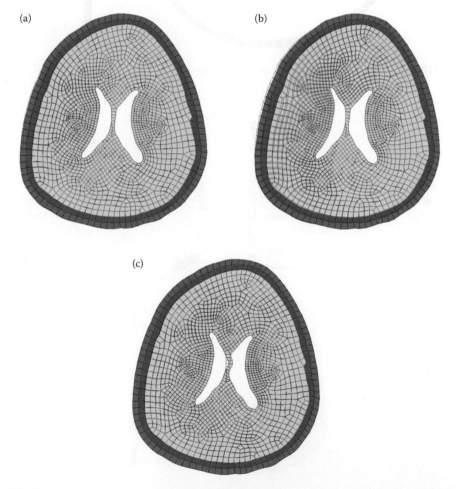

FIGURE 11.3 Deformation of the head at different steps. (a) Step 1 $Dmax = 0.643\mathrm{e}{-}3\mathrm{m}$; (b) Step 2 $Dmax = 1.259\mathrm{e}{-}3\mathrm{m}$; (c) Step 3 $Dmax = 2.115\mathrm{e}{-}3\mathrm{m}$.

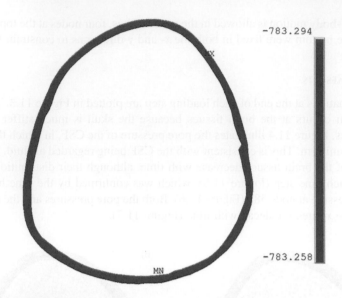

FIGURE 11.4   Pore pressures of CSF at the last step (Pa).

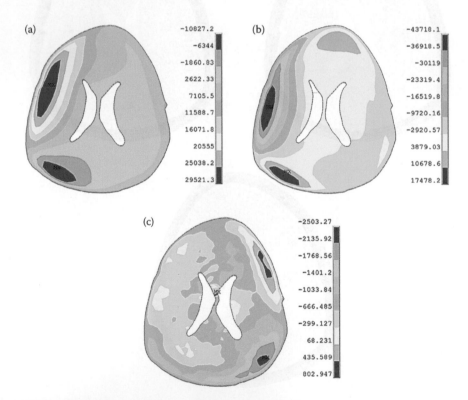

FIGURE 11.5   Pore pressures of brain tissues at the various steps (Pa). (a) Pore pressures at step 1; (b) Pore pressures at step 2; (c) Pore pressures at step 3.

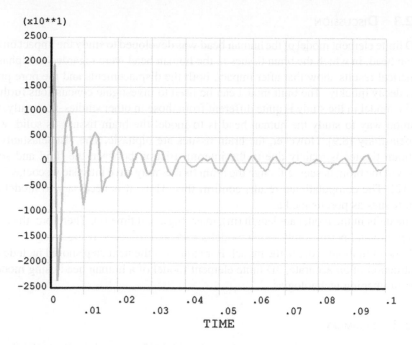

FIGURE 11.6   Time history of pore pressures of the brain tissues.

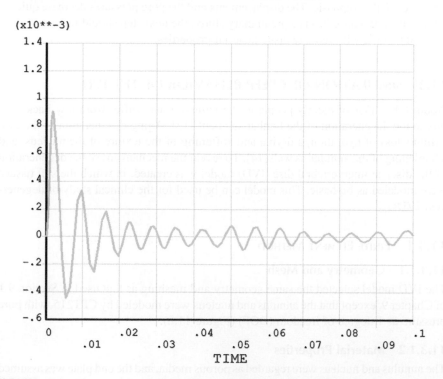

FIGURE 11.7   Time history of displacement of node 82 in the *x*-direction.

11.2.3 DISCUSSION

A 2D finite element model of the human head was developed to study the impact on the human head, in which the brain tissues in the human head were regarded as biphasic. Numerical results show that after impact, both the displacements and the pore pressures decay quickly. The built model can be used to investigate concussions further.

The model in the study is quite different from those in other studies. Currently, the common way to study the human head is to model the brain tissues as solid, with viscoelasticity [8,9]. However, the brain tissues are biphasic. The viscoelasticity of the brain tissues is mainly due to the internal friction between the fluid and solid phases of the brain tissues, although the brain tissues may have intrinsic viscoelasticity [10–12]. The computational results confirm this. Thus, it is necessary to model the brain tissues as porous media.

The units in the model are length (m), mass (kg), and time (s). Therefore, the units of pressure and permeability in the model are Pa and $m^4/N/s$, respectively.

Some limitations exist in the model. For example, the next step should include the construction of an accurate, 3D finite element model of a human head using modern medical imaging technologies.

11.2.4 SUMMARY

The impact on the head was simulated in ANSYS, in which the brain tissues were modeled as biphasic. The displacements and the pore pressures decrease quickly after impact. Because this is a preliminary study, the next step should focus on building a 3D model with more realistic material properties.

11.3 SIMULATION OF CREEP BEHAVIOR OF THE IVD

About 60%–85% of elderly people in the United States suffer from low back pain due to the degeneration of the lumbar segment [13]. Aging degeneration of the disc exhibits loss of hydration, a drying and stiffening of the texture of the nucleus, and a hardening of the annulus as well [14]. To reveal the mechanisms of the degeneration of the disc, an intervertebral disc (IVD) model was created, in which the soft tissues were modeled as biphasic. This model can be used for the clinical study of degenerated IVDs.

11.3.1 FINITE ELEMENT METHOD

11.3.1.1 Geometry and Mesh

The IVD model selected the same geometry and meshing as that used in Section 9.1 of Chapter 9, except that the annulus and nucleus were modeled by CPT215, with pore pressure as 1 degree of freedom (DOF) (Figure 11.8).

11.3.1.2 Material Properties

The annulus and nucleus were regarded as porous media, and the end plate was assumed only as solid. The material properties of these elements are listed in Table 11.2 [15].

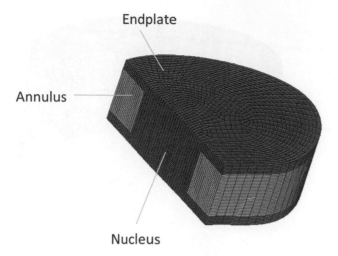

Endplate

Annulus

Nucleus

FIGURE 11.8 Finite element model of half an IVD.

TABLE 11.2
Material Properties of the IVD Parts

	Bone	Nucleus	Annulus
Young's modulus (MPa)	223.8	0.75	1.5
Poisson's ratio	0.4	0.17	0.17
Permeability ($mm^4/N/s$)	N/A	1e–3	2e–4

11.3.1.3 Loading and Boundary Conditions

The lateral surfaces of the annulus are permeable. Thus, the pore pressures at the lateral surfaces were specified as zero. The top surface of the IVD was compressed by a force of 196 N, and the bottom was constrained with all DOFs. Since half an IVD was modeled, the symmetrical conditions were applied along the x-axis (Figure 11.9).

The following lists the APDL commands that define the pore pressure boundary conditions:

```
ASEL, S, AREA, , 4
NSLA, S, 1
NSEL, R, LOC, Z, 1.5, 9.5
D, ALL, PRES, 0              ! pore pressure zero
ALLSEL
```

11.3.1.4 Solution Setting

Stepped loading was applied at the beginning. The total testing time, 3.07e6 seconds (about 12.3 days), was examined to simulate the creep behavior of the IVD.

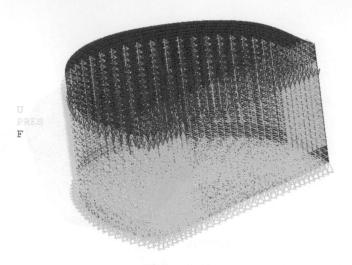

FIGURE 11.9 Boundary conditions of the IVD.

The full Newton-Raphson method with unsymmetrical matrices of elements was selected for CPT elements:

```
/SOLVE
TT = 3.07E6          ! total time
TIME, TT
! full Newton-Raphson method with unsymmetrical matrices of
  elements
NROPT, UNSYM
KBC, 1               ! stepped loading
NSUB, 50, 1000, 20
OUTRES, ALL, ALL
SOLVE
```

11.3.2   RESULTS

Figure 11.10 illustrates the final deformation of the IVD with a maximum displacement of 4.25 mm. The time history of the vertical displacement of the IVD appears in Figure 11.11, which indicates that the top surface of the IVD moves down gradually after the loading is applied during the first step; after 5 days, it approaches a constant value.

The pore pressures of the annulus and nucleus in the first step (which lasts 0.25 day) are illustrated in Figure 11.12a. After 12 days, the pore pressure dissipates (Figure 11.12b), which is confirmed by the time history of the pore pressure of node 10047 (Figure 11.13).

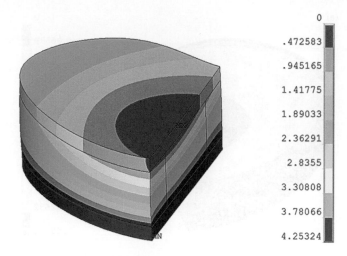

FIGURE 11.10   Deformation of the IVD at the last step (mm).

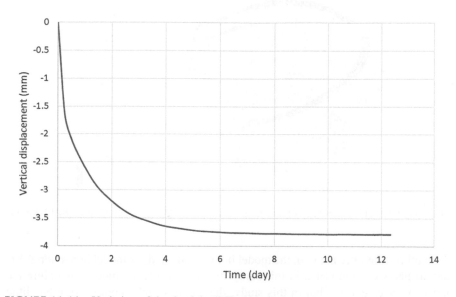

FIGURE 11.11   Variation of depth of the IVD with time.

11.3.3   DISCUSSION

In the developed IVD model, the annulus and nucleus were regarded as biphasic. Under compression loading, the top of the IVD moves down gradually, while the pore pressures dissipate. That is the creep response of the IVD due to friction between the fluid and solid phases in the IVD. If the soft tissues are modeled as only solid, as in Section 9.1, then the IVD would not exhibit the creep behavior.

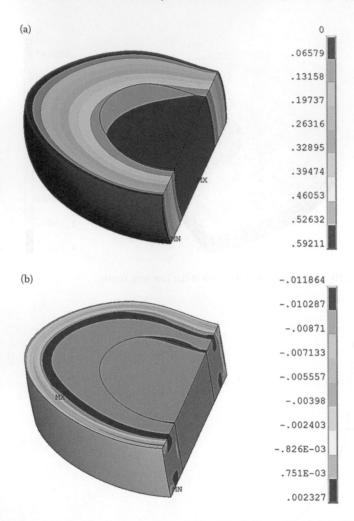

FIGURE 11.12   Pore pressures of soft tissues in the IVD (MPa). (a) 0.25 day; (b) 12.3 days.

Another difference between the model in this study and the model in Section 9.1 is that no fibers were modeled in this study. In Section 9.1, the model (including the fibers) is under tension, but in this study, the annulus under compression has little deformation in the radical direction, and little fibers within the annulus are in tension. As the fibers are only in tension, their presence or absence has no influence on the computational results of this study. Of course, if the compressive loading is very high and the annulus undergoes very large deformation in the radical direction, many fibers within the annulus will be in tension. Thus, it would be more appropriate to include fibers in the model.

This study is preliminary. Numerous studies have worked in this area, including looking at the biphasic theory, triphasic theory, poroelastic theory, nonlinear material, and dynamical state [16]. This work defines the direction of future research.

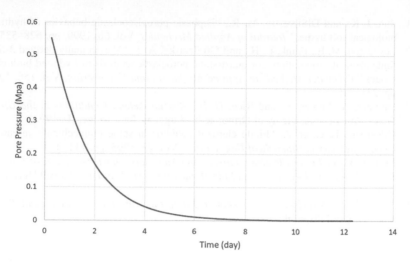

FIGURE 11.13   Time history of the pore pressure of node 10047.

11.3.4 SUMMARY

A poroelastic model of the IVD was created to simulate its creep response, which confirms the interaction of two phases in the soft tissues.

REFERENCES

1. ANSYS19.0 Help Documentation in the help page of product ANSYS190.
2. Vermeer, P. A., and Verrujit, A., "An accuracy condition for consolidation by finite elements." *International Journal for Numerical and Analytical Methods in Geomechanics*, Vol. 5, 1981, pp. 1–14.
3. Bruns, J. J., and Hauser, W. A., "The epidemiology of traumatic brain injury: A review." *Epilepsia*, Vol. 44, 2003, pp. 2–10.
4. Chan, H. S., "Mathematical model for closed head impact." *18th Stapp Car Crash Conference of the Society of Automotive Engineers*, Ann Arbor, MI, 1974, pp. 557–579.
5. Ward, C., "Finite element models of the head and their use in brain injury research." *26th Stapp Car Crash Conference of the Society of Automotive Engineers*, Ann Arbor, MI, 1982, pp. 71–85.
6. Hosey, R. R., and Liu, Y. K., "A homeomorphic finite element model of the human head and neck." *Finite Elements in Biomechanics*, 1982, pp. 379–401.
7. Ruan, J. S., Khalil, T., and King, A. I., "Dynamic response of the human head to impact by three-dimensional finite element analysis." *Journal of Biomechanical Engineering*, Vol. 116, 1994, pp. 44–51.
8. Chen, H. X., *Finite Element Investigation of Closed Head Injuries*. University of Manitoba, 2010.
9. Wang, C. Z., *Finite Element Modeling of Blast-Induced Traumatic Brain Injury*. University of Pittsburgh, 2013.
10. Suh, J. K., and Bai, S., "Finite element formulation of biphasic poroviscoelastic model for articular cartilage." *Journal of Biomechanical Engineering*, Vol. 120, 1998, pp. 195–201.

11. Suh, J. K., and DiSilvestro, M. R., "Biphasic poroviscoelastic behavior of hydrated biological soft tissue." *Journal of Applied Mechanics*, Vol. 66, 1999, pp. 528–535.
12. Levenston, M. E., Frank, E. H., and Grodzinsky, A. J., "Variationally derived 3-field finite element formulations for quasi-static poroelastic analysis of hydrated biological tissues." *Computer Methods in Applied Mechanics and Engineering*, Vol. 156, 1998, pp. 231–246.
13. Paremer, A., Fumer, S., and Rice, D. P., *Musculoskeletal Conditions in the United States*. American Academy of Orthopaedic Surgeons, Park Ridge, IL, 1992.
14. Gilbertson, L. G. et al., "Finite element methods in spine biomechanics research." *Critical Reviews in Biomedical Engineering*, Vol. 23, 1995, pp. 411–473.
15. Massey, C. J., *Finite Element Analysis and Materials Characterization of Changes Due to Aging and Degeneration of the Human Intervertebral Disc*. Drexel University, 2009.
16. Yang, Z. C., *Poroviscoelastic Dynamic Finite Element Model of Biological Tissue*. University of Pittsburgh, 2004.

Part III

Joints

Joints are the regions where two or more bones meet. They connect bones in the body and make them into a functional whole. Clinical studies demonstrate that some joint diseases are associated with the mechanical loadings and contact pressure in the joints. For example, osteoarthritis (OA) is linked to the contact pressures on the knee joint. Thus, Part III studies the joint simulation.

Chapter 12 introduced the structure and function of joints. Then, a three-dimensional (3D) whole-knee model and a two-dimensional (2D) axisymmetrical poroelastic knee model were built in Chapter 13. Chapter 14 implemented the discrete element method in ANSYS to analyze the contact pressure of the knee joint and overcome contact convergence difficulties.

12 Structure and Function of Joints

Joints are classified as three types according to the type of allowed motion: immovable, slightly movable, and movable. Immovable joints exist generally in the skull, where the bony plate ends (i.e., joints) are connected by fibrous or cartilaginous elements. Slightly movable joints connect the articulating surfaces using broad, flattened discs of fibrocartilage. Hyaline cartilage covers the bony portions of the joint, and a fibrous capsule covers the entire structure. Slightly movable joints are found between the vertebrae, the distal tibiofibular articulation, the public symphysis, and the uppermost parts of the sacroiliac joint.

Most of the joints of the extremities are movable (Figure 12.1). These joints, which have articular cartilage with an extremely low coefficient of friction, allow a wide range of motion [1]. The articulating bone surfaces are covered by a very thin layer of cortical bone (compact bone). Between the cortical bone and articular cartilage is an intermediate layer of the calcified cartilage. The articular cartilage is bonded to

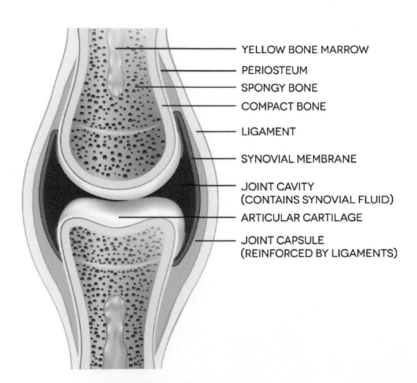

YELLOW BONE MARROW

PERIOSTEUM

SPONGY BONE

COMPACT BONE

LIGAMENT

SYNOVIAL MEMBRANE

JOINT CAVITY
(CONTAINS SYNOVIAL FLUID)

ARTICULAR CARTILAGE

JOINT CAPSULE
(REINFORCED BY LIGAMENTS)

FIGURE 12.1   Knee joint. (guniita © 123RF.com.)

the bony end plate and surrounded by a set of collagen fibers. The hyaline articular cartilage is characterized as a set of smooth and resilient connective tissues and serves as the bearing and gliding surfaces. The joint cavity contains a thin layer of synovial fluid. This structure provides almost frictionless mobility, which is essential to the joint function.

REFERENCE

1. Wooley, P. H., Grimm, M. J., and Radin, E. L., "The structure and function of joints." *Arthritis & Allied Conditions*, 15th edition, Lippincott Williams & Wilkins, Philadelphia, 2005, pp. 151–152.

13 Modeling Contact

This chapter focuses on modeling knee contact. Section 13.1 introduces the three contact types in biology. Then, Sections 13.2 and 13.3 model knee contacts as solid and porous media, respectively.

13.1 CONTACT MODELS

Contacts exist everywhere in biology, especially in the joints. The most common contact is surface to surface. Thus, the contact pair can be modeled by TARGE169/CONTA171 in two-dimensional (2D) problems and TARGE170/CONTA174 or TARGE170/CONTA175 in three-dimensional (3D) problems [1]. Because bone is much stiffer than cartilage, the bones in a contact pair with cartilages are always modeled by TARGE169/TARGE170, and the cartilages in the contact pair are defined by CONTA171/CONTA174/CONTA175.

In general, the contact problems are represented by adding the constraint equations to the regular governing equations. Different contact behaviors correspond to different constraint equations. The common contact behaviors include standard contact, rough frictional contact (i.e., no sliding), bonded contact, no separation contact (i.e., sliding is permitted), and internal multipoint constraint. Regarding the contact behavior, contacts in biology are separated into three categories:

1. *Standard contact*: Contact between cartilages and the meniscus in the joints is considered as standard contact (Figure 13.1), in which separation and sliding are allowed between the contact pair.
2. *Always bonded*: One of the examples of being always bonded is contact between the femur and femoral cartilage, which is defined by keyopt(12) = 5 (Figure 13.1). Another example is the connection between ligaments and bones (Figure 13.1).
3. *Multipoint constraint (MPC)*: Compared to soft tissues, bones are regarded as rigid; a pilot node (Figure 13.1), which is implemented by MPC with keyopt(2) = 2, controls their rigid motion.

Besides the selection of appropriate contact elements and contact behaviors, completing contact problems requires paying close attention to the following points:

- Verify the normal directions of the contact and target surfaces in the correct directions; if not, reverse them.
- When too much penetration affects the convergence, increase the FKN value to reduce the contact penetration values.
- If the contact does not work, check the pinball region to see if the PINB constant is too small.

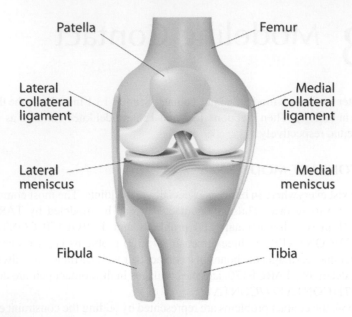

FIGURE 13.1 Knee joint. (designua © 123RF.com.)

13.2 3D KNEE CONTACT MODEL

Osteoarthritis (OA) is a joint disease characterized by degeneration of the articular cartilage and subchondral bone (Figure 13.2). Approximately 27 million adults in the United States struggle with OA [2,3]. Studies show that the health and function of the joint are directly related to the joint's mechanical loadings, and contact pressure in the knee joint structure is a crucial factor that causes knee pain and OA [4]. To effectively understand and appropriately assess knee joint biomechanics, the finite element method has been applied to study knee contact [5–8]. In this chapter, the knee contact was simulated in ANSYS.

13.2.1 Finite Element Model

13.2.1.1 Geometry and Mesh

Finite element analysis was performed on the knee model from Open Knee [9] (under the terms of the Creative Commons Attribution 3.0 License). The study focused on the contacts among the meniscus, tibia, femur, and their cartilages. It was assumed that only an external load was applied to the femur to restrict movements of the knee joint. Therefore, ligaments and tendons, as well as patella, were removed from the model. The whole finite element model consists of the femur and its cartilage, the tibia and its cartilage, and the meniscus (Figure 13.3), comprising 60,533 elements and 96,853 nodes. Compared to cartilage, bone is considered as rigid. To reduce computational time, only the part of the bone connected to the cartilages was modeled (Figure 13.4).

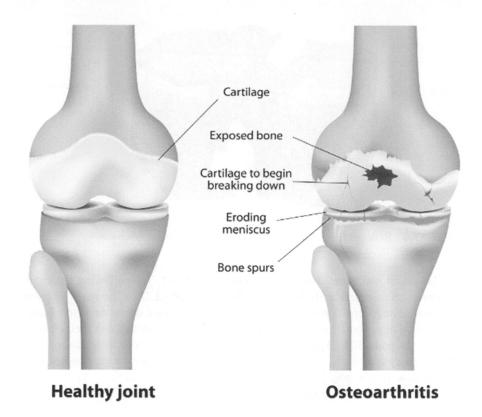

Healthy joint   **Osteoarthritis**

FIGURE 13.2   Schematic of OA. (designua © 123RF.com.)

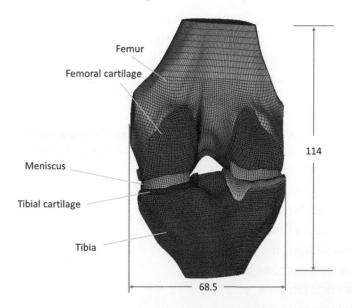

FIGURE 13.3   Finite element model of a knee joint (all dimensions in millimeters).

FIGURE 13.4   Reduced finite element model of the knee joint for computation.

In the model, the bones were meshed with SHELL181. The following algorithm was developed to generate 3D SOLID185 from 2D SHELL181. The basic idea was to find the normal direction of each shell element, and then create four new nodes in the normal direction. The four new nodes and the four nodes of one shell element form an eight-node SOLID185 element. The newly created nodes were applied with MPC boundary conditions, and those nodes in the shell elements were coupled with the cartilages. That is why 3D bone elements are required.

The following lists the commands that convert the four-node SHELL181 to the eight-node Solid185:

```
ET, 3, 185
*GET, NUM_E, ELEM, 0, COUNT        ! get number of elements
*GET, E_MIN, ELEM, 0, NUM, MIN     ! get min element number
*DIM, E1, ARRAY, NUM_E, 4
*DO, I, 1, NUM_E, 1                 ! output to ASCII by looping over
                                      elements
CURR_E = E_MIN
P1X = NX(NELEM(CURR_E, 1))          ! get the coordinates of three
                                      nodes in shell element
P1Y = NY(NELEM(CURR_E, 1))
P1Z = NZ(NELEM(CURR_E, 1))
P2X = NX(NELEM(CURR_E, 2))
P2Y = NY(NELEM(CURR_E, 2))
P2Z = NZ(NELEM(CURR_E, 2))
P3X = NX(NELEM(CURR_E, 3))
P3Y = NY(NELEM(CURR_E, 3))
P3Z = NZ(NELEM(CURR_E, 3))
```

```
! calculate the normal direction of shell elements
PX = (P2Y - P1Y) * (P3Z - P2Z) - (P2Z - P1Z) * (P3Y - P2Y)
PY = (P2Z - P1Z) * (P3X - P2X) - (P2X - P1X) * (P3Z-P2Z)
PZ = (P2X - P1X) * (P3Y - P2Y) - (P2Y - P1Y) * (P3X - P2X)
PP = SQRT(PX * PX + PY * PY + PZ * PZ)
PXX = PX/PP
PYY = PY/PP
PZZ = PZ/PP
*DO, J, 1, 4
XX = NX(NELEM(CURR_E,J)) - PXX
YY = NY(NELEM(CURR_E,J)) - PYY
ZZ = NZ(NELEM(CURR_E,J)) - PZZ
N, A + (I - 1) * 4 + J, XX, YY, ZZ    ! create new nodes along the
                                        normal direction
*ENDDO
TYPE, 3
MAT, 5
! create new eight-node Solid185 element
E, NELEM(CURR_E,1), NELEM(CURR_E,2), NELEM(CURR_E,3), NELEM
(CURR_E,4), A+(I-1)*4+1, A+(I-1)*4+2, A+(I-1)*4+3, A+(I-1)
*4+4
*GET, E_MIN, ELEM, CURR_E, NXTH
*ENDDO
```

13.2.1.2   Material Properties

All materials in the model were simplified to be linear isotropic. Their material properties are listed in Table 13.1 [10].

13.2.1.3   Contact Pairs

Six contact pairs exist in the knee model. All of them belong to a surface-to-surface contact, which is modeled by TARGE170 and CONTA175. Based on the contact behavior, they are separated into three categories.

TABLE 13.1

Material Properties of the Knee Joint

	Young's Modulus (MPa)	Poisson's Ratio
Bone	20e3	0.3
Cartilage	5	0.46
Meniscus	59	0.49

(a)

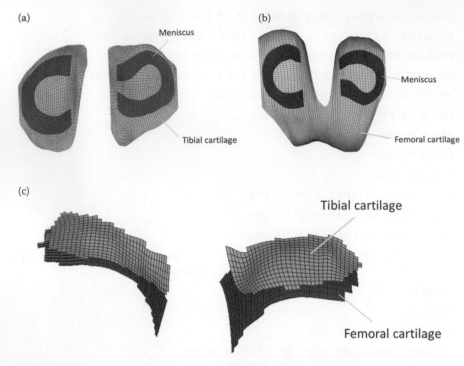

(b)

(c)

FIGURE 13.5   Standard contacts in the knee joint. (a) Meniscus versus tibial cartilage; (b) meniscus versus femur cartilage; (c) femoral cartilage versus tibial cartilage.

1. The standard contact category consists of the femoral cartilage–meniscus contact, tibial articular cartilage–meniscus contact, and femoral cartilage–tibial articular cartilage contact (Figure 13.5). Examples of the commands to define contact between the tibial articular cartilage and meniscus are listed here:

```
! contact between tibial articular cartilage and meniscus
MAT, 2
R, 3
REAL, 3
ET, 7, 170
ET, 8, 175

KEYOPT, 8, 9, 1
KEYOPT, 8, 10, 0
KEYOPT, 8, 12, 0
KEYOPT, 8, 5, 2
R, 3,
! generate the target surface
```

```
CMSEL, S, TCS, NODE
TYPE, 7
ESLN, S, 0
ESURF
CMSEL, S, ELEMCM
! generate the contact surface
CMSEL, S MMTIBR, NODE
CMSEL, A, LMTIBR, NODE
TYPE, 8
ESLN, S, 0
ESURF
```

2. In the always bonded contact, the cartilages of the knee are always bonded to the bones (as the name implies). Thus, the tibia—tibial articular cartilage contact and femur—femoral cartilage contact were assumed to be always bonded and defined by keyopt(12) = 5 in the contact pair (Figure 13.6). The commands for the contact between the tibia and tibial articular cartilage are presented here:

```
! contact between tibia and tibial articular cartilage
MAT, 2
R, 5
REAL, 5
ET, 11, 170
ET, 12, 175
KEYOPT, 12, 9, 1
KEYOPT, 12, 12, 5                    ! bonded (always)
```

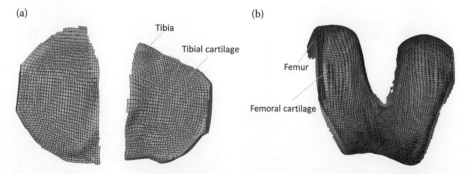

FIGURE 13.6   Contacts between cartilages and bones. (a) Tibia and tibial cartilage; (b) femur and femoral cartilage.

```
KEYOPT, 12, 5, 3                    ! close gap/reduce penetration
R, 5,
! generate the target surface
CMSEL, S, INTCS, NODE
TYPE, 11
ESLN, S, 0
ESURF
CMSEL, S, _ELEMCM
! generate the contact surface
CMSEL, S, TC2TIB, NODE
TYPE, 12
ESLN, S, 0
ESURF
ALLSEL
```

3. In the MPC rigid motion, the bone is about 1,000 times stiffer than the cartilage. Compared to the cartilage, bone is considered as rigid, and its motion
   is de fined by an internal multiple point constraint (MPC) using keyopt(2) =
   2 (Figure 13.7). Here is one example:

```
MAT, 4
R, 15
REAL, 15
ET, 17, 170
ET, 18, 175
KEYOPT, 18, 12, 5                   ! always bonded
```

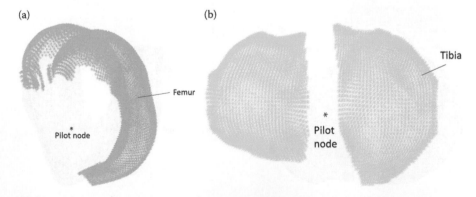

(a)   (b)

Femur

Tibia

Pilot node

Pilot
node

FIGURE 13.7 MPC constraints. (a) Coupling between a pilot node and femur; (b) coupling
between a pilot node and tibia.

```
KEYOPT, 18, 4, 2
KEYOPT, 18, 2, 2                          ! MPC
KEYOPT, 17, 2, 1
KEYOPT, 17, 4, 111111                     ! all DOFs are constrained
TYPE, 17
! create a pilot node
N, 10000001, 80, 55, 55
TSHAP, PILO
E, 10000001
TYPE, 18
! generate the contact surface
NSEL, S, NODE, , 4E6, 5E6
ESLN, S, 0
ESURF
```

13.2.1.4 Boundary Conditions

To simulate the mechanical behavior of a knee when the tibia is fixed and the femur is moved down 1 mm vertically, the corresponding conditions were applied to the pilots because the pilot nodes control the bone's rigid motions. In addition, the ends of the meniscus attached to the tibia were constrained in all degrees of freedom (DOFs).

	.117494
	3.43634
	6.75518
	10.074
	13.3929
	16.7117
	20.0306
	23.3494
	26.6682
	29.9871

FIGURE 13.8 von Mises stresses of the meniscus (MPa).

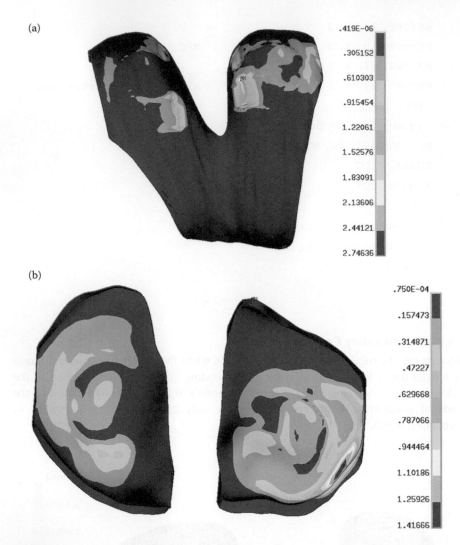

FIGURE 13.9   von Mises stresses of cartilages (MPa). (a) Femoral cartilage; (b) tibial cartilage.

13.2.2 RESULTS

Figures 13.8 through 13.10 plot the von Mises stresses of the meniscus, tibia, and femur, as well as their cartilages. The stresses of cartilages have two causes. One is mainly from contact between the meniscus and cartilages, and the other comes from the contact between the cartilages. The tibial cartilage and femoral cartilage have maximum stresses of 1.42 and 2.75 MPa, respectively (much less than that of the meniscus's 30 MPa). The bone has a similar stress distribution as the cartilages, although the maximum stresses of the tibia and femur are 1.92 and 3.21 MPa, respectively. The contact pressures of contact pairs appear in Figures 13.11 through 13.13.

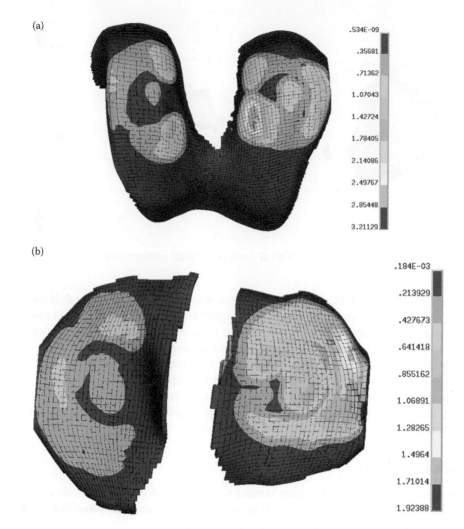

FIGURE 13.10 von Mises stresses of the bone (MPa). (a) Femur; (b) tibia.

The contact pressures of the meniscus are localized in a small area with high value. On the contrary, the contact pressures between the cartilages distributed much more evenly. The contact pressures between the cartilages and bones were similar to the stresses of the cartilages and bones because their contacts were defined as always bonded.

13.2.3 DISCUSSION

The finite element model of a full knee was built with six contact pairs defined. Based on their contact behavior, the six contact pairs were defined as standard contact,

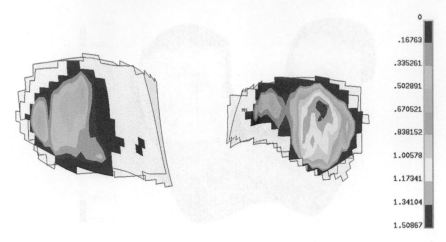

	0
	.16763
	.335261
	.502891
	.670521
	.838152
	1.00578
	1.17341
	1.34104
	1.50867

FIGURE 13.11　Contact pressure of tibial cartilage–femoral cartilage contact (MPa).

always bonded, and MPC, respectively. Solution convergence was reached; because the obtained contact pressure for each contact pair looked reasonable, the built model was partially validated. To validate the model fully, a comparison between the computational results and the experimental data is necessary.

Some limitations exist in the current model. The simple linear elastic material definition does not match the real materials of the knee, especially the meniscus that was demonstrated as anisotropic. Furthermore, to replicate the knee's mechanical response fully in daily life, the force loading in the stance phase of the gait should be selected to simulate a dynamic process instead of static loading.

13.2.4　SUMMARY

The finite element analysis of knee contact was conducted in ANSYS, with various contact pairs defined. The nonlinear solution converged. Although some limitations exist, the model can be used to study the knee contact.

13.3　2D POROELASTIC MODEL OF KNEE

In Section 13.2, the meniscus and cartilages in the knee model were presented only as solid. However, as introduced in Chapter 6, cartilages and the meniscus are porous and 60%–85% fluid. According to biphasic theory [11], the fluid phase of soft tissues carries most of the load in a physiologically relevant, short loading time. Therefore, to fully understand the true physiological biomechanical behavior of the soft tissues of the knee joint, the soft tissues of the knee joint should be modeled as porous media. Therefore, a 2D, axisymmetrical finite element model of the knee was developed in ANSYS, in which the soft tissues were modeled by coupled pore-pressure thermal (CPT) elements.

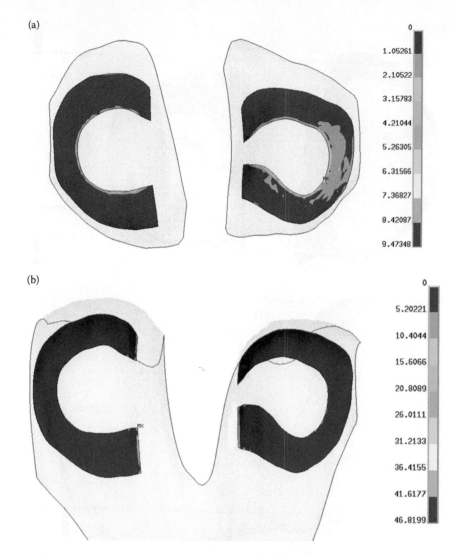

FIGURE 13.12 Contact pressure of cartilages—meniscus contact (MPa). (a) Tibial cartilage—meniscus contact pressure; (b) femoral cartilage—meniscus contact pressure.

13.3.1 FINITE ELEMENT MODEL

13.3.1.1 Geometry and Mesh

A 2D, axisymmetric tibial–femoral cartilage layer model was created to simulate the knee contact (Figure 13.14). The tibial articular cartilage was simplified as a rectangular area attached to the bone, and the femoral cartilage was modeled by a curve tangent to the top surface of the meniscus and tibial cartilage. The bottom surface of the

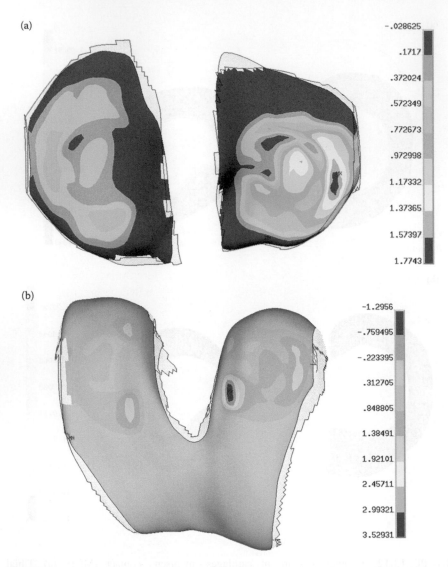

(a)

```
 −.028625
    .1717
  .372024
  .572349
  .772673
  .972998
  1.17332
  1.37365
  1.57397
  1.7743
```

(b)

```
 −1.2956
 −.759495
 −.223395
  .312705
  .848805
  1.38491
  1.92101
  2.45711
  2.99321
  3.52931
```

FIGURE 13.13 Contact pressure of cartilages—bone contact (MPa). (a) Tibia–tibial cartilage; (b) femur–femoral cartilage.

meniscus was assumed to be flat. The upper surface of the meniscus was cubic and tangent to the tibial cartilage, while its outer surface was quadratic. The top of the model included part of the femur. The model was implemented in ANSYS, in which the bone was meshed by PLANE182 and the cartilages and the meniscus were meshed by CPT212.

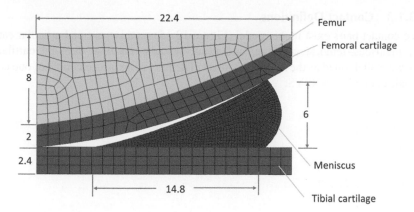

FIGURE 13.14 2D axisymmetrical finite element model of a knee joint (all dimensions in millimeters).

13.3.1.2 Material Properties

Both bone and cartilage were assumed to be linear elastic. The cartilages were modeled as porous media. The meniscus was regarded as porous media, with the solid phase as orthotropic material [12–17]:

```
! bone material
MP, EX, 1, 17600
MP, NUXY, 1, 0.3
! cartilage
FPX2 = 3              ! units mm⁴/N/s
MP, EX, 2, 0.69,   ! MPa
MP, NUXY, 2, 0.18,
TB, PM, 2,,, PERM
TBDATA, 1, FPX2, FPX2, FPX2
! meniscus
FPX3 = 1.26          ! units mm⁴/N/s
MP, EX, 3, 0.075   ! MPa ! Young's modulus in radial direction
MP, EY, 3, 0.075   ! Young's modulus in vertical direction
MP, EZ, 3, 100     ! Young's modulus in circumferential direction
MP, NUXY, 3, 0.18,
MP, NUXZ, 3, 0.45
MP, NUYZ, 3, 0.45
TB, PM, 3,,, PERM
TBDATA, 1, FPX3, FPX3, FPX3
```

13.3.1.3   Contact Definitions

Three contact pairs exist in the model (Figure 13.15): meniscus–tibial articular cartilage, meniscus–femoral cartilage, and tibial articular cartilage–femoral cartilage.
They were all defined as the standard contact. The following is an example of one contact pair defined in ANSYS:

```
ET, 4, 169
ET, 5, 171
R, 5

KEYOPT, 5, 12, 0
KEYOPT, 5, 1, 8

! generate the target surface
REAL, 4
TYPE, 4
NSEL, S, , , N2
ESURF
ALLSEL
! generate the contact surface
TYPE, 5
NSEL, S, , , N4
ESURF
ALLSEL
```

13.3.1.4   Boundary Conditions and Loading

The top surface of the femur was loaded with nodal force of 147 N. The bottom surface of the model was fixed in all DOFs (Figure 13.16). The pore pressures at the outer
surfaces of the cartilages and meniscus were assumed to be zero.

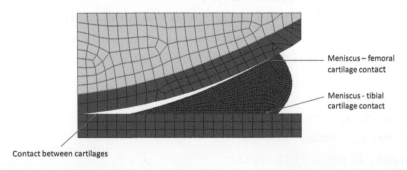

Meniscus – femoral
cartilage contact

Meniscus - tibial
cartilage contact

Contact between cartilages

FIGURE 13.15   Contacts in the knee model.

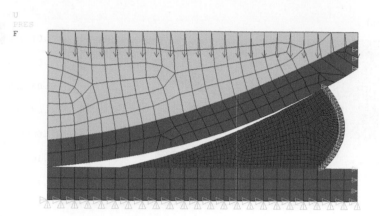

FIGURE 13.16   Boundary conditions and loadings.

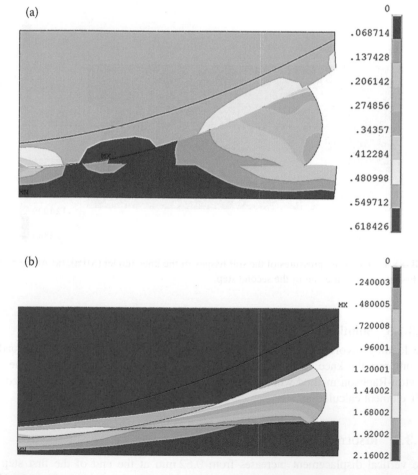

FIGURE 13.17   Deformation of the knee (mm). (a) At the end of the first step; (b) at the end of the second step.

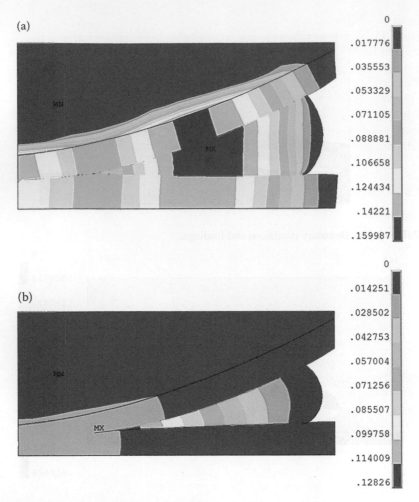

FIGURE 13.18   Pore pressures of the soft tissues in the knee model (MPa). (a) At the end of the first step; (b) at the end of the second step.

13.3.1.5   Solution Setting

The loading involves two steps. In the first step, the external forces were loaded to compress the knee. The loading was held for 120 s in the second step. The full Newton-Raphson method with unsymmetrical matrices of elements was defined for CPT element calculation.

13.3.2   Results

The vertical displacement increases from 0.62 mm at the end of the first step to 2.16 mm at the end of the second step (Figure 13.17). During this time, the pore pressures drop significantly due to water being expelled from the cartilages

(Figure 13.18). That is consistent with the time history of the nodal pore pressure (Figure 13.19).

Figure 13.20 shows the stresses of the meniscus in the circumferential direction at the end of the two steps, which indicate that the biggest stress occurred at the outside of the meniscus.

The contact stress distribution is much different at periods of 1 s and 120 s (Figure 13.21). At 1 s, the contact pressures of the three contact pairs are not very different. However, at 120 s, the contact pressure between the cartilages increases, with significant differences from that of the other two contact pairs.

13.3.3 DISCUSSION

When the soft tissues are modeled as porous media, the vertical displacement under compression moves down with time and approaches a constant, while the water is expelled. In addition, the porous pressures reduce with time. Porous media modeling also influences the contact pressures of the various contact pairs. If the soft tissues are considered only as solid, these changes will not happen.

The meniscus is anisotropic, and its Young's modulus in the circumferential direction is about 1,000 times higher than that in the other two directions. Thus, its stress in the circumferential direction takes over most of the loadings. The computational results show that the biggest stress of the meniscus occurs at the outside of the meniscus, which looks reasonable because the outside part has the biggest strain in the circumferential direction.

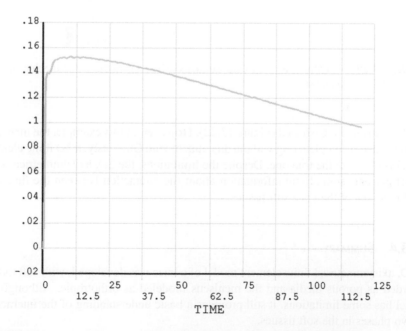

FIGURE 13.19 Pore pressures of the soft tissues in the knee model with time.

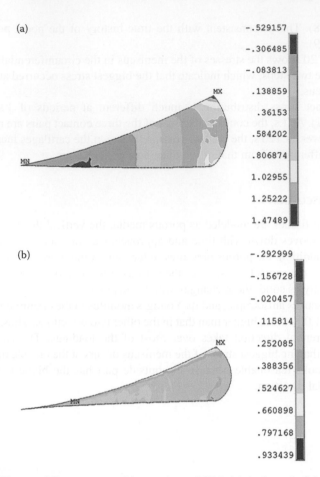

(a)

-.529157
-.306485
-.083813
.138859
.36153
.584202
.806874
1.02955
1.25222
1.47489

(b)

-.292999
-.156728
-.020457
.115814
.252085
.388356
.524627
.660898
.797168
.933439

FIGURE 13.20 von Mises stresses of the meniscus (MPa). (a) At the end of the first step; (b) at the end of the second step.

The menisci are C-shaped (Figure 13.22). However, in this example, the meniscus is modeled as a complete ring without any attachment. Obviously, a 3D finite element model is closer to the true one. Despite the limitations, the 2D, axisymmetrical knee model presents some basic information about the interaction between the fluid and solid matrixes of the knee soft tissues.

13.3.4 SUMMARY

A 2D, axisymmetrical finite element model of a knee was developed, with soft tissues regarded as porous media and the meniscus modeled as orthotropic. Although the model has some limitations, it still provides a basic understanding of the interaction of two phases in the soft tissues.

(a)

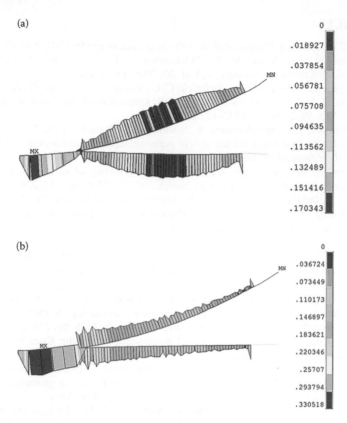

(b)

FIGURE 13.21 Contact pressures of the knee model (MPa). (a) At the end of the first step; (b) at the end of the second step.

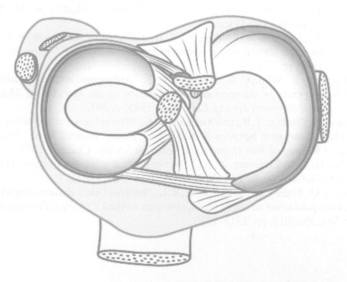

FIGURE 13.22 Schematic of a knee joint. (Aksana Kulchytskaya ©123RF.com.)

REFERENCES

1. ANSYS 190 Help Documentation in the help page of product ANSYS190.
2. Arden, N., and Nevitt, M. C., "Osteoarthritis: Epidemiology." *Best Practice & Research Clinical Rheumatology*, Vol. 20, 2006, pp. 3–25.
3. Murphy, L., Schwartz, T. A., Helmick, C. G., Renner, J. B., Tudor, G., Koch, G., and Jordan, J. M., "Lifetime risk of symptomatic knee osteoarthritis." *Arthritis & Rheumatism*, Vol. 59, 2008, pp. 1207–1213.
4. Andriacchi, T. P., Mündermann, A., Smith, R. L., Alexander, E. J., Dyrby, C. O., and Koo, S., "A framework for the *in vivo* pathomechanics of osteoarthritis at the knee." *Annals of Biomedical Engineering*, Vol. 32, 2004, pp. 447–457.
5. Taylor, Z. A., and Miller, K., "Constitutive modeling of cartilaginous tissues: A review." *Journal of Applied Biomechanics*, Vol. 22, 2006, pp. 212–229.
6. Peña, E., Del Palomar, A. P., Calvo, B., Martínez, M., and Doblaré, M., "Computational modelling of diarthrodial joints. Physiological, pathological, and post-surgery simulations." *Archives of Computational Methods in Engineering*, Vol. 14, 2007, pp. 47–91.
7. Kazemi, M., Dabiri, Y., and Li, L. P., "Recent advances in computational mechanics of the human knee joint." *Computational and Mathematical Methods in Medicine*, 2013, pp. 1–27.
8. Mononen, M. E., Tanska, P., Isaksson, H., and Korhonen, R. K., "A novel method to simulate the progression of collagen degeneration of cartilage in the knee: Data from the osteoarthritis initiative." *Scientific Reports*, Vol. 6, 2016, pp. 214–215.
9. Open Knee(s): Virtual Biomechanical Representations of the Knee Joint. Website: simtk.org/projects/openknee.
10. Kumar, V. A. and Jayanthy, A. K., "Finite element analysis of normal tibiofemoral joint and knee osteoarthritis: A comparison study validated through geometrical measurements." *Indian Journal of Science and Technology*, Vol. 9, 2016.
11. Mow, V. C., Kuei, S. C., Lai, W. M., and Armstrong, C. G., "Biphasic creep and stress relaxation of articular cartilage in compression, theory and experiments." *Journal of Biomechanical Engineering*, Vol. 102, 1980, pp. 73–84.
12. Chern, K. Y., Zhu, W. B., Kelly, M. A., and Mow, V. C., "Anisotropic shear properties of bovine meniscus." Transaction of the 36th Annual Meeting of the Orthopaedic Research Society, New Orleans, 1990, pp. 246.
13. Proctor, C., Schmidt, M. B., Whipple, R. R., Kelly, M. A., and Mow, V. C., "Material properties of the normal medial bovine meniscus." *Journal of Orthopaedic Research*, Vol. 7, 1989, pp. 771–782.
14. Whipple, R. R., Wirth, C. R., and Mow, V. C., "Anisotropic and zonal variations in the tensile properties of the meniscus." *Transaction of the 31st Annual Meeting of the Orthopaedic Research Society*, Las Vegas, 1985, p. 367.
15. Cohen, B., Gardner, T. R., and Ateshian, G. A., "The influence of transverse isotropy on cartilage indentation behavior-A study of the human humeral head." *Transaction of Annual Meeting of 39th Orthopaedic Research Society*, Chicago, 1993, p. 185.
16. Reilly, D. T., and Burstein, A. H., "The mechanical properties of cortical bone." *Journal of Bone & Joint Surgery*, Vol. 56, 1974, pp. 1001–1022.
17. Guo, H. Q., Maher, S. A., and Spilker, R. L., "Biphasic finite element contact analysis of the knee joint using an augmented Lagrangian method." *Medical Engineering & Physics*, Vol. 35, 2013, pp. 1313–1320.

14 Application of the Discrete Element Method for Study of the Knee Joint

Contact problems are always challenging. They may have convergence difficulties that make it impossible to complete the computation. Thus, discrete element analysis (DEA) has been developed to solve contact problems, which will be discussed in this chapter.

14.1 INTRODUCTION OF DISCRETE ELEMENT METHOD

Discrete element analysis (DEA) is an alternative to solve joint contact problems [1–2]. Its basic idea is to build tons of springs between the bones to replace the cartilages and to solve the simple linear equations to acquire the contact pressure (Figure 14.1).

In the joint contact problem, DEA is based on the following assumptions:

- Cartilages are assumed to be linear and elastic.
- The size of the contact area is significantly larger than the thickness through the cartilage.
- Subchondral bone is regarded as rigid.

DEA methods are more efficient than those of finite element analysis (FEA), as the contact is simplified in order to solve some linear equations; yet, the DEA methods are limited in order to approximate the joint contact pressures between two rigid bodies.

DEA methods have been applied to the hip [3–5], the ankle [6], the knee joint [7–9], and the patellofemoral [10–12]. This chapter focuses on the knee joint using DEA in ANSYS.

14.2 FINITE ELEMENT MODEL

The Discrete Element Model only requires the bone surfaces in the computation. Thus, the cartilages and meniscus were removed from the Finite Element Model of the knee (Section 13.2), and the whole model was comprised of the tibia and femur. The next step is to build the springs between the bones to represent the soft tissues. First, however, it is important to introduce the mathematical theory needed to calculate the line-plane intersection.

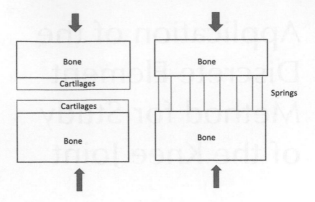

FIGURE 14.1   Schematic of the Discrete Element Method.

14.2.1   LINE-PLANE INTERSECTION

Line **AB**, formed by points **A** and **B**, crosses a plane determined by points **P0**, **P1**, and **P2** (Figure 14.2). The intersection point **C** is expressed by [13]

$$\mathbf{C} = \mathbf{A} + \mathbf{AB} \cdot t \tag{14.1}$$

$$t = \frac{(\mathbf{P}_{01} \times \mathbf{P}_{02}) \cdot (\mathbf{A} - \mathbf{P}_0)}{-\mathbf{AB} \cdot (\mathbf{P}_{01} \times \mathbf{P}_{02})} \tag{14.2}$$

$$u = \frac{(\mathbf{P}_{02} \times -\mathbf{AB}) \cdot (\mathbf{A} - \mathbf{P}_0)}{-\mathbf{AB} \cdot (\mathbf{P}_{01} \times \mathbf{P}_{02})} \tag{14.3}$$

$$v = \frac{(-\mathbf{AB} \times \mathbf{P}_{01}) \cdot (\mathbf{A} - \mathbf{P}_0)}{-\mathbf{AB} \cdot (\mathbf{P}_{01} \times \mathbf{P}_{02})} \tag{14.4}$$

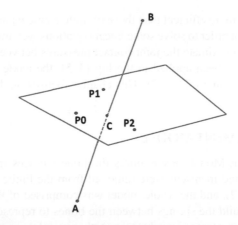

FIGURE 14.2   Schematic of line-plane intersection. Line AB crosses a plane formed by three points at point C.

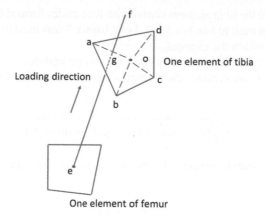

FIGURE 14.3   Schematic of building a spring element *eg*.

where u and v are used to calculate the location of point **C**. If $u + v \leq 1$ is satisfied, then the intersection point **C** lies in the triangle formed by the three points **P0**, **P1**, and **P2**.

14.2.2   Building Springs

Building springs between the tibia and the femur involves the following steps:

1. For one element of the femur, find its center *e* and form line *ef* along the loading direction. Line *ef* should be long enough to cross the tibia.
2. Each element of the tibia is divided into four triangles formed by the tibia's element center *o* with its four nodes (Figure 14.3).
3. A line-plane intersection calculation of line *ef* with each triangle created in step 2 following Eqs. (14.1) through (14.4) is done.
4. For all elements of the tibia, steps 1–3 are repeated until the intersection point *g* lies in the triangle ($u + v \leq 1$). Then points *g* and *e* are used to create one spring. Note: Point *e* is the center of the femur element, and point *g* is the new tibia node created by the line-plane intersection method.
5. Steps 1–4 for all elements of the femur are repeated to build all springs (Figure 14.4).

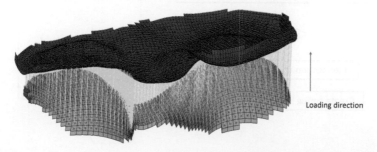

FIGURE 14.4   Springs between bones. The springs are aligned with the loading direction.

In step 2, before the tibia element center with four nodes formed the four triangles, the same algorithm used in Section 3.2.2.4 of Chapter 3 was used to verify if the tibia element center is within the element.

Figure 14.5 presents the entire flowchart for this procedure.

After the springs are created, their lengths are calculated from the nodal coordinates:

$$l_{eg} = \sqrt{(e_x - g_x)^2 + (e_y - g_y)^2 + (e_z - g_z)^2}. \tag{14.5}$$

Assuming the initial length of spring l_0 as 6.5 mm, the initial strain is computed by

$$e_0 = \frac{(l_{eg} - l_0)}{l_0}. \tag{14.6}$$

The initial strain is defined in ANSYS by

```
INIS, SET, CSYS, -2        ! follow the element coordinate system
INIS, SET, DTYP, EPEL
INIS, DEFINE , , , , , e_0
```

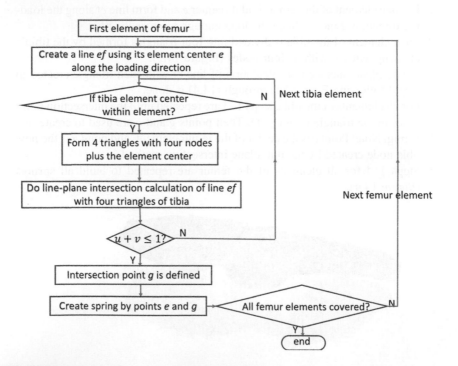

FIGURE 14.5 Flowchart to build springs between bones.

In addition, the springs are regarded as compression only, and their cross-sectional areas are equal to those of the femur elements. These are specified in the section definitions:

```
SECDATA, C1(I,1)  ! c1(i,1) are areas of the femur elements
obtained by *GET
SECCONTROL, , -1   ! compression only
```

Appendix 14 lists the entire set of ANSYS Parametric Design Language (APDL) commands that define springs.

14.2.3 BOUNDARY CONDITIONS

The bones were meshed with SHELL181. To display the stress contour of the femur, three-dimensional (3D) SOLID185 elements were created from two-dimensional (2D) SHELL181 following the same approach described in Section 13.2.1.1 of Chapter 13. The newly created nodes of the femur at the bottom were constrained in all degrees of freedom (DOFs). All nodes of the tibia were applied with displacement of -1 mm in the z-direction to represent its rigid motion.

14.2.4 RESULTS

Figure 14.6 illustrates the von Mises stresses of the femur, which approximate those in Figure 13.10 in Section 13.2.2 if the central stresses caused by the meniscus contact are ignored. This indicates that the developed DEA algorithm can be used to estimate the contact pressures in the knee joint.

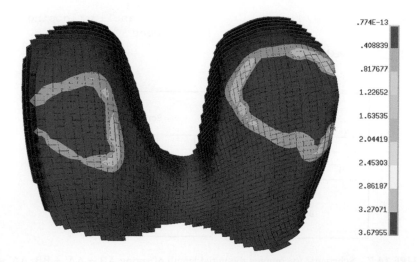

FIGURE 14.6   von Mises stresses of the femur (MPa).

14.2.5 DISCUSSION

The Discrete Element Method was implemented to study the joint contact. The method is based on one assumption—that the change of relative positions between cartilages primarily determines the contact pressure. As a result, the contact between the bones can be simulated by a cluster of springs.

In the simulation, the initial length of the springs was assumed as 6.5 mm uniformly, which is different from the real-life case. To overcome this discrepancy, the thickness of cartilage can be calculated using the line-plane algorithm. Thus, the initial length of one spring is easily defined (Figure 14.7).

In this example, the loading direction is in the z-direction of the global coordinates. When the loading direction was not aligned with the axis of the global coordinate, a local coordinate system $o'x'y'z'$ was created, in which the loading direction is in the direction of the $o'x'$-axis. The translation matrix of the local system **R** is expressed as

$$\mathbf{R} = \begin{bmatrix} R_{11} & R_{12} & R_{13} \\ R_{21} & R_{22} & R_{23} \\ R_{31} & R_{32} & R_{33} \end{bmatrix}. \tag{14.7}$$

Each row in the **R** matrix refers to the directions of vectors $o'x'$, $o'y'$, and $o'z'$, respectively.

The following APDL codes were developed to modify the nodal coordinates to make the global coordinates align with the loading direction:

```
NSLE, S
*GET, NUM_N, NODE, 0, COUNT      ! get number of nodes
*GET, N_MIN, NODE, 0, NUM, MIN ! get min node number
SHPP, OFF
*DO, I, 1, NUM_N, 1              ! output to ascii by looping
                                ! over nodes

CURR_N = N_MIN
```

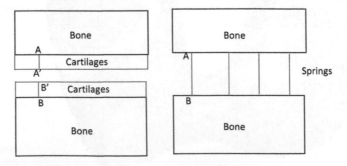

FIGURE 14.7 Schematic to compute the initial length of spring AB = AA' + BB. AA' and BB' are obtained using the line-plane intersection algorithm.

```
XNEW = R11*NX(CURR_N) + R12*NY(CURR_N) + R13*NZ(CURR_N)

YNEW = R21*NX(CURR_N) + R22*NY(CURR_N) + R23*NZ(CURR_N)

ZNEW = R31*NX(CURR_N) + R32*NY(CURR_N) + R33*NZ(CURR_N)

NMODIF, CURR_N, XNEW, YNEW, ZNEW

*GET, N_MIN, NODE, CURR_N, NXTH

*ENDDO
```

Overall, DEA is a fast method to estimate the contact pressure of cartilages. In cases where the traditional contact has a convergence issue, it offers a viable alternative.

14.2.6 SUMMARY

The DEA algorithm was developed to study the knee contact using the line-plane intersection method. The results are close to that of the standard contact method, which indicate that this new method is feasible for studying the knee contact.

REFERENCES

1. Li, G., Sakamoto, M., and Chao, E. Y., "A comparison of different methods in predicting static pressure distribution in articulating joints." *Journal of Biomechanics*, Vol. 30, 1997, pp. 635–638.
2. An, K. N., Himeno, S., Tsumura, H., Kawai, T., and Chao, E. Y., "Pressure distribution on articular surfaces: Application to joint stability evaluation." *Journal of Biomechanics*, Vol. 23, 1990, pp. 1013–1020.
3. Yoshida, H., Faust, A., Wilckens, J., Kitagawa, M., Fetto, J., and Chao, E. Y., "Three-dimensional dynamic hip contact area and pressure distribution during activities of daily living." *Journal of Biomechanics*, Vol. 39, 2006, pp. 1996–2004.
4. Genda, E., Iwasaki, N., Li, G., MacWilliams, B. A., Barrance, P. J., and Chao, E. Y., "Normal hip joint contact pressure distribution in single-leg standing—Eeffect of gender and anatomic parameters." *Journal of Biomechanics*, Vol. 34, 2001, pp. 895–905.
5. Abraham, C. L., Maas, S. A., Weiss, J. A., Ellis, B. J., Peters, C. L., and Anderson, A. E., "A new discrete element analysis method for predicting hip joint contact stresses." *Journal of Biomechanics*, Vol. 46, 2013, pp. 1121–1127.
6. Kern, A. M., and Anderson, D. D., "Expedited patient-specific assessment of contact stress exposure in the ankle joint following definitive articular fracture reduction." *Journal of Biomechanics*, Vol. 48, 2015, pp. 3427–3432.
7. Anderson, D. D., Iyer, K. S., Segal, N. A., Lynch, J. A., and Brown, T. D., "Implementation of discrete element analysis for subject-specific, population-wide investigations of habitual contact stress exposure." *Journal of Applied Biomechanics*, Vol. 26, 2010, pp. 215–223.
8. Segal, N. A., Anderson, D. D., Iyer, K. S., Baker, J., Torner, J. C., Lynch, J. A., and Brown, T. D., "Baseline articular contact stress levels predict incident symptomatic knee osteoarthritis development in the MOST cohort." *Journal of Orthopaedic Research*, Vol. 27, 2009, pp. 1562–1568.
9. Halloran, J. P., Easley, S. K., Petrella, A. J., and Rullkoetter, P. J., "Comparison of deformable and elastic foundation finite element simulations for predicting knee

replacement mechanics." *Journal of Biomechanical Engineering*, Vol. 127, 2005, pp. 813–818.

10. Elias, J. J., and Saranathan, A., "Discrete element analysis for characterizing the patellofemoral pressure distribution: Model evaluation." *Journal of Biomechanical Engineering*, Vol. 135, 2013, pp. 81011.

11. Elias, J. J., Wilson, D. R., Adamson, R., and Cosgarea, A. J., "Evaluation of a computational model used to predict the patellofemoral contact pressure distribution." *Journal of Biomechanics*, Vol. 37, 2004, pp. 295–81302.

12. Smith, C. R., Won Choi, K., Negrut, D., and Thelen, D. G., "Efficient computation of cartilage contact pressures within dynamic simulations of movement." *Computer Methods in Biomechanics and Biomedical Engineering: Imaging & Visualization*, Vol. 1, 2016, pp. 1–8.

13. Foley, J., Dam, A. V., Feiner, S., and Hughes, J., "Clipping Lines," *Computer Graphics*. 3rd ed, Addison–Wesley, Boston, 2013.

Part IV

Simulation of Implants

Each year, millions of patients undergo surgical procedures that involve implanted medical devices designed to improve their quality of life. The term *implant* refers to a device that replaces or acts as a fraction of an entire biological structure. Currently, implants are being applied in many different parts of the body for applications such as orthopedics, pacemakers, and cardiovascular stents. The big demand for implanted medical devices sets a high standard for them, requiring a better mechanical design and more reliable materials. Finite element analysis has been proven to be a very useful tool in the investigation and optimization of the design of implanted medical devices, also providing novel insights about wear and fatigue fracture mechanics.

Chapter 15 simulated the contact between the talar component and the bone in the ankle replacement, and discussed how the material properties of the bone affect the contact. Stent implantation was modeled using the shape memory alloy (SMA) material model to obtain the deformation and stresses of the stent described in Chapter 16. In Chapter 17, wear of the liner of the hip implant was simulated using the Archard wear model. Finally, Chapter 18 used Separating, Morphing, Adaptive, and Remeshing Technology (SMART) in ANSYS190 to study the fatigue crack growth of a mini dental implant (MDI) for prediction of its fatigue life.

15 Study of Contact in Ankle Replacement

In 1992, the U.S. Food and Drug Administration (FDA) approved the Agility Total Ankle Replacement System [1]. Since then, both the United States and Canada have widely used this system (Figure 15.1). However, clinical studies found that medial tilt emerges as one of the two big concerns in total ankle arthroplasty [2] and is associated with the contact between the bones and the talar component. One study indicates that the medial tilt is due to the variety of material properties across the bone [3]. With the aid of the TBFIELD command described in Section 3.2 in Chapter 3, this study focuses on the contact between the bones and the talar component and tries to reveal the relation between the medial tilt and the material properties of the bone.

15.1 FINITE ELEMENT MODEL

15.1.1 Geometry and Mesh

This section focuses on the contact between the bones and the talar components. This model is comprised of the bones and talar component. The talar component was meshed with 10 node tetrahedral elements. The talus is comprised of the cortical and cancellous bones. The cortical bone was meshed by SHELL181 with a thickness of 1.8 mm (Figure 15.2), while the cancellous bone was meshed with tetrahedral elements (Figure 15.2). The whole model comprises 41,355 elements and 68,213 nodes.

15.1.2 Material Properties

The Young's modulus and Poisson's ratio of the talar component were defined as 193 GPa [4] and 0.3, respectively.

The bone included both the cortical bone and cancellous bone, and the cortical bone is a layer with a thickness of 1.8 mm on the surface of the talus (Figure 15.2). The material properties of the cortical bone were simplified as isotropic material with a Young's modulus of 17,600 MPa and Poisson's ratio of 0.3 [5].

As discussed in Section 3.2, cancellous bone is essentially nonhomogeneous; its material properties were interpolated from Jensen's experimental data using the TBFIELD command [6]. Three interpolation methods are available in ANSYS190: the Linear Multivariate (LMUL), the Radial-Basis (RBAS), and the Nearest-Neighbor (NNEI). In this study, only 36 experimental points were involved for interpolation. The RBAS was selected for interpolation because (1) it does not require too much computation and (2) it is a global method that ensures C1 continuity. The material definition by the RBAS is listed in Section 3.2.

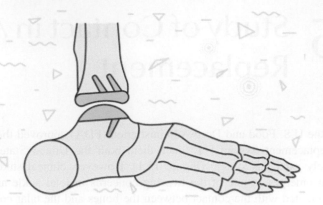

FIGURE 15.1   Schematic of ankle replacement. (Shopplaywood ©123RF.com.)

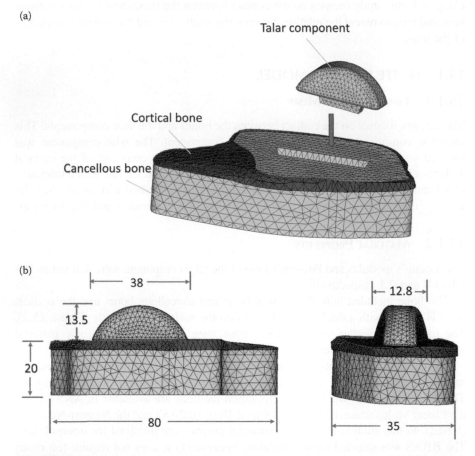

FIGURE 15.2   Finite element model of contact between the talar component and the bone. (a) contact between the talar component and the bone; (b) the dimensions of the bone and talar component (mm).

Under compression, the cancellous bone also yields at strains that approximate 1% and fails at around 2% [7]. The cancellous bone should be defined by a plastic model; otherwise, the obtained stresses may be far beyond the yield stress of the bone. For simplification, the cancellous bones are defined as having perfect plasticity, with a yield strain at 1.5% [3]. The perfect plasticity was interpolated by the RBAS method as follows:

```
FF=0.015
TB,PLASTIC,5,,,MISO
TBFIELD, XCOR, 191.75
TBFIELD, YCOR, -22.951
TBFIELD, ZCOR, 135
TBPT,DEFI, 0, 111 * FF          ! yield point
TBPT,DEFI, 1E-2, 111 * FF       ! perfect plasticity
......
TBFIELD, XCOR, 140.79
TBFIELD, YCOR, -35.642
TBFIELD, ZCOR, 145
TBPT,DEFI, 0, 222 * FF          ! yield point
TBPT,DEFI, 1E-2, 222 * FF       ! perfect plasticity
TBIN, ALGO, RBAS
```

15.1.3 Contact Definition

Because bone ingrowth bonds the talus to the implant in current designs, the bone-implant interface was modeled as being perfectly bonded (Figure 15.3).

15.1.4 Loading and Boundary Conditions

This study assumed that 1,500 N of force was applied on the surface of the talar component, which is twice that of a body weight of 75 kg [8]. Because the loading area is

FIGURE 15.3 Contact pairs between the bone and the talar component.

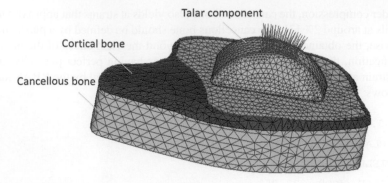

FIGURE 15.4 Loading on the talar component.

about 100 mm², the pressure loaded is specified as 15 MPa. The loading direction is vertical to simulate the vertical standing case (Figure 15.4).

The bottom surface of the talus was fixed in all six degrees of freedom (DOFs).

15.2 RESULTS

The displacements of the cancellous bone and the associated elemental von Mises stresses, as well as the elemental von Mises plastic strains when the forces are loaded in the vertical direction, appear in Figures 15.5 through 15.7, respectively. Figure 15.5 clearly shows that the deformation of the cancellous bone is not uniform, but instead tilts in one direction. Obviously, the stress distribution is not uniform either. The major plastic strain occurs at the contact area, with the maximum plastic strain of 0.60 at the anterior edge.

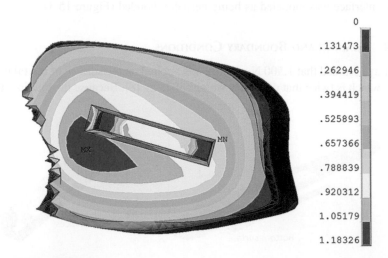

0	
.131473	
.262946	
.394419	
.525893	
.657366	
.788839	
.920312	
1.05179	
1.18326	

FIGURE 15.5 Deformation of the bone (mm).

.049471
.986857
1.92424
2.86163
3.79902
4.7364
5.67379
6.61117
7.54856
8.48595

FIGURE 15.6 von Mises stresses of the bone (MPa).

15.3 DISCUSSION

It is quite understandable that the medial tilting of the component takes place when the Young's modulus distribution (Section 3.2) is considered. The tilting occurs from the high Young's modulus to the low Young's modulus. The location where the largest deformation transpires has the lowest Young's modulus. These results indicate that a variety of Young's moduli of the cancellous bones cause the medial tilting of the component.

The big stresses occurred at the edge of the talar component, where it reaches plasticity and its stresses equal the yield stress. Because the yield stress is assumed to be

0
.066432
.132864
.199297
.265729
.332161
.398593
.465025
.531457
.59789

FIGURE 15.7 Plastic strains of the bone.

proportional to the Young's modulus, the big stresses happened at the location with the large Young's modulus.

The big plastic strains occur at the edge of the talar component. In this study, the perfect bonded contact was assumed to be between the talar component and the bone. However, the talar component's Young's modulus is about 1,000 times higher than that of the bone. The talar component is regarded as the rigid body. Thus, the contact area of the bone experiences continuous strain. The big strains, including plastic strains, always occur at the edge of the contact area when the stresses reach the yield stresses.

No direct evidence supports the bond assumption between the bone and the talar component. Therefore, it would be more appropriate to model the contact as the standard contact with a high coefficient of friction.

15.4   SUMMARY

A three-dimensional (3D) finite element model of the Agility Total Ankle Replacement System was built to study the contact between the bone and the talar component. In this model, the material of the cancellous bones in the talus was assumed to have perfect plasticity using tb, plastic, MISO, and the material properties were interpolated from the experimental data to represent the variation of the material properties across the talus using tbin, algo, RBAS. The computational results indicate that the medial tilting of the component is due to a variety of the Young's moduli of the cancellous bones.

REFERENCES

1. Gougoulias, N. E., Khanna, A., and Maffulli, N., "History and evolution of total ankle arthroplasty." *British Medical Bulletin*, Vol. 89, 2009, pp. 111–151.
2. Hintermann, B., Total ankle arthroplasty: Historical overview, current concepts and future perspectives. Springer, Austria, 2005.
3. Cui, Y., Hu, P., Wei, N., Cheng, X., Chang, W., and Chen, W., "Finite element study of implant subsidence and medial tilt in agility ankle replacement." *Medical Scientific Monitor*, Vol. 24, 2018, pp. 1124–1131.
4. Miller, M. C., Smolinski, P., Conti, S., and Galik K., "Stresses in polyethylene liners in a semiconstrained ankle prosthesis." *Journal of Biomechanical Engineering*, Vol. 126, 2004, pp. 636–640.
5. Rho, J. Y., Kuhn-Spearing, L., and Zioupos, P., "Mechanical properties and the hierarchical structure of bone." *Medical Engineering & Physics*, Vol. 20, 1998, pp. 92–102.
6. Jensen, N. C., Hvid, I., and Krøner, K., "Strength pattern of cancellous bone at the ankle joint." *Engineering in Medicine*, Vol. 17, 1988, pp. 71–76.
7. Kopperdahl, D. L., and Keaveny, T. M., "Yield strain behavior of trabecular bone." *Journal of Biomechanics*, Vol. 31, 1998, pp. 601–608.
8. Stauffer, R. N., Chao, E. Y. S., and Brewster, R. C., "Force and motion analysis of the normal, diseased, and prosthetic ankle joint." *Clinical Orthopaedics and Related Research*, Vol. 127, 1977, pp. 189–196.

16 Simulation of Shape Memory Alloy (SMA) Cardiovascular Stent

The shape memory alloy (SMA) is a special alloy that "remembers" its original shape after deformation, either through elastic recovery after great deformation or by returning to its initial shape when heated. It has two unique effects that, in general, do not exist in traditional materials. The first effect—the superelastic effect—(Figure 16.1a) returns to the original shape of SMA after great deformation. The second—the shape memory effect—(Figure 16.1b) removes the permanent deformation after the temperature increases. The unique character of SMA makes it widely used in the field of biomedical engineering, such as eyeglass frames, dental wires, and vascular stents.

16.1 SMA MODELS

Two SMA material models are available to simulate these behaviors: the SMA model for superelasticity [1] and SMA material model with the shape memory effect [2–4].

16.1.1 SMA MODEL FOR SUPERELASTICITY

The stress σ–strain ε relation of SMA for superelasticity is given by [1]

$$\sigma = \mathbf{D}{:}(\varepsilon - \varepsilon_{tr}), \tag{16.1}$$

where $\mathbf{D}$ is the elastic stiffness tenor and ε_{tr} is the transformation strain tensor.

The phase transformation is governed by the Drucker-Prager function as follows (Figure 16.2):

$$F = q + 3\alpha p, \tag{16.2}$$

where

$$q = \sqrt{\frac{3}{2}\mathbf{S}{:}\mathbf{S}} \tag{16.3}$$

$$\mathbf{S} = \sigma - p\mathbf{1} \tag{16.4}$$

$$p = \frac{1}{3}\sigma{:}\mathbf{1}. \tag{16.5}$$

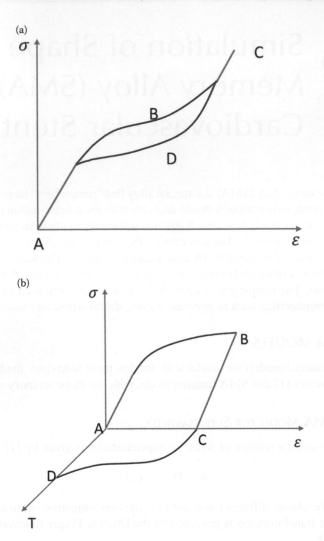

FIGURE 16.1 Two SMA features. (a) SMA model for superelasticity; (b) SMA material model with the shape memory effect.

Here, α is the material parameter used to measure the material behaviors in tension and compression.

Thus, the evolution of the martensite fraction, ξ_s, is expressed as a function of F:

$$
\dot{\xi}_s =
\begin{cases}
-H^{AS}(1 - \xi_s)\dfrac{\dot{F}}{F - R_f^{AS}} & A \rightarrow S \\[4mm]
H^{SA}\xi_s\dfrac{\dot{F}}{F - R_f^{SA}} & S \rightarrow A
\end{cases},
\tag{16.6}
$$

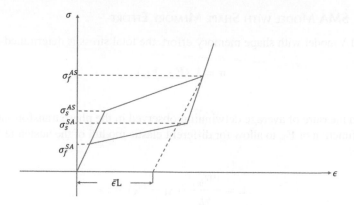

FIGURE 16.2 Material parameters to define SMA superelasticity.: σ_s^{AS} and σ_f^{AS} are starting and final stress values for the forward phase transformation, respectively. σ_s^{SA} and σ_f^{SA} are starting and final stress values for the reverse phase transformation, respectively. $\bar{\varepsilon}_L$ is maximum residual strain.

where

$$R_f^{AS} = \sigma_f^{AS}(1 + \alpha) \tag{16.7}$$

$$R_f^{SA} = \sigma_f^{SA}(1 + \alpha) \tag{16.8}$$

$$H^{AS} = \begin{cases} 1 & \text{if } R_s^{AS} < F < R_f^{AS} \quad \text{and} \quad \dot{F} > 0 \\ 0 & \text{otherwise} \end{cases} \tag{16.9}$$

$$H^{SA} = \begin{cases} 1 & \text{if } R_f^{SA} < F < R_s^{SA} \quad \text{and} \quad \dot{F} < 0 \\ 0 & \text{otherwise} \end{cases} \tag{16.10}$$

$$R_s^{AS} = \sigma_s^{AS}(1 + \alpha) \tag{16.11}$$

$$R_s^{SA} = \sigma_s^{SA}(1 + \alpha) \tag{16.12}$$

Therefore, the transformation strain tensor, ε_{tr}, is determined by

$$\dot{\varepsilon}_{tr} = \dot{\xi}_s \bar{\varepsilon}_L \frac{\partial F}{\partial \sigma}. \tag{16.13}$$

Here, $\bar{\varepsilon}_L$ is the material parameter shown in Figure 16.2.

Eqs. (16.1) through (16.13) indicate that in order to define the SMA model for superelasticity, one needs to input these material parameters: σ_s^{AS}, σ_f^{AS}, σ_s^{SA}, σ_f^{SA}, $\bar{\varepsilon}_L$, and α. Therefore, the SMA model for superelasticity is defined in ANSYS by the following commands:

```
TB, SMA, , , , SUPE
TBDATA, 1, σₛᴬˢ, σ_f^AS, σₛˢᴬ, σ_f^SA, ēₗ, α.
```

16.1.2 SMA MODEL WITH SHAPE MEMORY EFFORT

In the SMA model with shape memory effort, the total stress is determined by [2–4]

$$\boldsymbol{\sigma} = \mathbb{L}{:}(\mathbf{E} - \mathbf{E}_{in}), \tag{16.14}$$

where

$\mathbf{E}_{in}$ is a measure of average detwinning observed in the phase transformation; and $\mathbb{L}$ is a function of $\mathbf{E}_{in}$ to allow for different elastic moduli of the austenite and martensite.

$$\mathbb{L} = \frac{||\mathbf{E}_{in}||}{\epsilon_L}(\mathbb{L}_{\mathbf{M}} - \mathbb{L}_{\mathbf{A}}) + \mathbb{L}_{\mathbf{A}} \tag{16.15}$$

When the material is in its austenite phase, $\mathbb{L} = \mathbb{L}_{\mathbf{A}}$; when the material is in its martensite phase, $\mathbb{L} = \mathbb{L}_{\mathbf{M}}$.

The yield function is assumed to be a Prager-Lode type in order to take asymmetric behavior of SMA in tension and compression into account:

$$F(\mathbf{X}_{tr}) = \sqrt{2J_2} + m\frac{J_3}{J_2} - R, \tag{16.16}$$

where

$$J_2 = \frac{1}{2}\left(\mathbf{X}_{tr}^2{:}\mathbf{1}\right) \tag{16.17}$$

$$J_3 = \frac{1}{3}\left(\mathbf{X}_{tr}^3{:}\mathbf{1}\right) \tag{16.18}$$

$$\mathbf{X}_{tr} = \boldsymbol{\sigma}' - [\tau_M(T) + h||\mathbf{E}_{in}|| + \gamma]\frac{\mathbf{E}_{in}}{||\mathbf{E}_{in}||} \tag{16.19}$$

$$\tau_M(T) = \langle \beta(T - T_0) \rangle^+. \tag{16.20}$$

Further, β is a material parameter, T is room temperature, T_0 is the temperature below which no twinned martensite occurs, and h is a material parameter associated with the hardening of the material during the phase transformation;

Here, γ is given by

$$\begin{cases} \gamma = 0 & \text{if } 0 < ||\mathbf{E}_{in}|| < \varepsilon_L \\ \gamma \geq 0 & \text{if } ||\mathbf{E}_{in}|| = \varepsilon_L \end{cases}, \tag{16.21}$$

here ε_L is a material parameter, m is Lode dependency parameter, and R is the elastic domain radius.

The evolution of $\mathbf{E}_{in}$ is related to the yield function F:

$$\mathbf{E_{in}} = \mathbf{E_{in}}(n) + \Delta\xi\frac{\partial F}{\partial\boldsymbol{\sigma}}. \tag{16.22}$$

Eqs (16.14) through (16.22) show that in order to define the SMA model with the shape memory effect, the input of seven material parameters is required, including h, T_0, R, β, ε_L, Em, and m. Thus, the SMA model with the shape memory effect is specified in ANSYS by the following commands:

```
TB,SMA,,,,MEFF
TBDATA, 1, h, T₀, R, β, εₗ
TBDATA,6, Em, m
```

16.2   SIMULATION OF ANGIOPLASTY WITH VASCULAR STENTING

Cardiovascular disease (CVD) is the major cause of death in the United States. The costs associated with CVD were estimated to be over \$430 billion in 2007. One of the most common and serious forms of CVD is atherosclerosis, which is characterized by the accumulation of material within an artery to reduce or block blood flow. A common and widely used treatment for atherosclerotic coronary arteries is to deploy an intravascular tubular prosthesis—namely, a stent (Figure 16.3). In the United States, 1.2 million people undergo stent implantations each year. Among these patients, nearly one-third of them require further intervention within half a year to reopen previously stented arteries [6,7]. The major issue linked to stenting

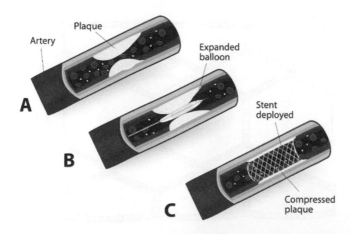

FIGURE 16.3   Schematic of angioplasty with vascular stenting. (Designua © 123RF.com.)

is restenosis. A sound knowledge of the mechanical behavior of stents helps to identify restenosis and contributes to the improvement of the design of stents. Thus, a three-dimensional (3D) finite element model was developed to study the stent implantation.

16.2.1 FINITE ELEMENT MODEL

16.2.1.1 Geometry and Mesh

The vessel was modeled with a straight cylinder with a length of 28 mm, inner diameter of 3.9 mm, and outer diameter of 4.3 mm. The thickness of the 8 mm-long plaque attached on the vessel varies along its length. Its thickness increases gradually from zero at the two ends to 0.35 mm in the middle (Figure 16.4). A small portion of a stent was selected for meshing and used in the model (Figure 16.5). The whole model, which was meshed by SOLID187, is comprised of 23,947 elements and 29,947 nodes.

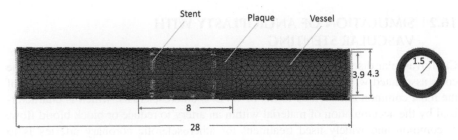

FIGURE 16.4　Finite element model of the stent implantation (all dimensions in millimeters).

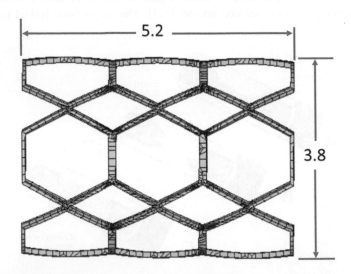

FIGURE 16.5　Finite element model of the stent (all dimensions in millimeters).

16.2.1.2 Material Properties

Angioplasty with vascular stenting only involves the superelasticity of SMA; no temperature change is required. Thus, the SMA model with superelasticity was selected for modeling the stent. The material properties of the stent in the form of SMA with the shape memory effect are described [5]:

```
MP, EX, 1, 53000
MP, NUXY, 1, 0.33
! define SMA material properties
TB, SMA, 1, , ,MEFF
TBDATA, 1, 1000, 280, 100, 6.1, 0.056, 53000
TBDATA, 7, 0.05
```

A simple uniaxial simulation with these MEFF material properties was performed on one SOLID185 element and obtained its strain-stress relation, as plotted in Figure 16.6. Compared to Figure 16.2, the material model for SMA SUPE is defined as

```
MP, EX, 1, 53000
MP, NUXY, 1, 0.33
! define SMA material properties
```

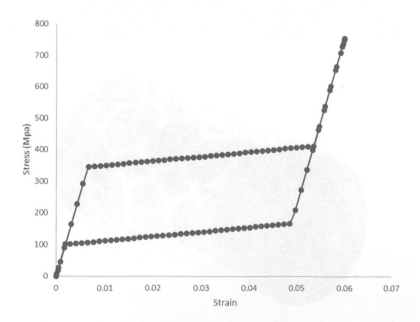

FIGURE 16.6 Strain–stress curve of SMA model.

```
TB, SMA, 1, , ,SUPE
TBDATA, 1, 345.2, 403, 168, 101, 0.056, 0
```

The vessel was modeled as Mooney-Rivlin material [8]:

```
C10 = 18.90E-3
C01 = 2.75E-3
C20 = 590.43E-3
C11 = 857.2E-3
NU1 = 0.49
DD=2 * (1 - 2*NU1)/(C10 + C01)


TB, HYPER, 2, ,5, MOONEY
TBDATA, 1, C10, C01, C20, C11, , DD
```

The plaque was assumed to be isotropic, with a Young's modulus of 2.19 MPa and a Poisson's ratio of 0.39.

16.2.1.3 Contact Pairs

The contact between the plaque and stent was defined as a standard contact with TARGE170 and CONTA174 (Figure 16.7). The contact key options are selected default. The contact was deactivated and activated as described in Section 16.2.1.4.

To allow radial expansion without rigid body motion, force-distributed boundary constraints were applied on two sides of the artery and stent. The definitions for these four contact pairs are very similar to the definitions used in Section 7.2 of Chapter 7 (Figure 16.8).

FIGURE 16.7 Contact between the plaque and the stent.

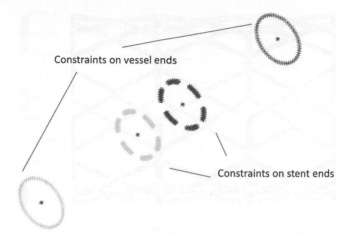

FIGURE 16.8   MPC constraints on the ends of the vessel and the stent.

16.2.1.4   Solution Setting
The simulation has three loading steps:

1. Deactivate the contact between the stent and plaque, and then apply pressure 0.15 MPa on the surface of the plaque and 0.04 MPa on the surface of the vessel to make the plaque and vessel expand past the radius of the stent.
2. Activate the contact between the stent and plaque.
3. Reduce the applying pressure to the normal blood pressure 0.0133 MPa (100 HHMg).

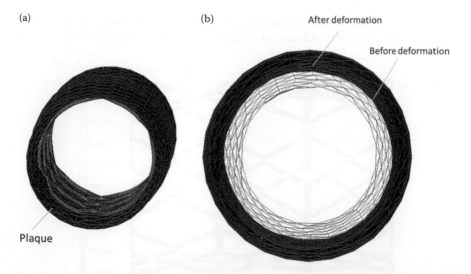

FIGURE 16.9   Deformation of the plaque. (a) Plaque and stent after deformation; (b) plaque before and after deformation.

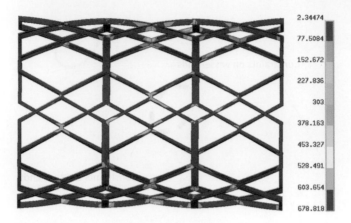

FIGURE 16.10   von Mises stresses of the stent (MPa).

16.2.2   RESULTS

The final deformation of plaque (Figure 16.9) indicates that the stent made the plaque expand from an initial radius of 1.5 mm to one of 1.9 mm, which allowed the stent surgery to achieve its goal of having more blood flow into the vessel. The von Mises stresses of the stent in Figure 16.10 have a peak value of 624 MPa, which is significantly beyond σ_s^{AS} (345 MPa). Thus, plastic deformation occurs in the stent with a maximum plastic strain of 0.039 (Figure 16.11). Compared to the stent, the plaque has much smaller stresses (Figure 16.12), with a peak value of 0.59 MPa.

16.2.3   DISCUSSION

The 3D finite element model was developed to study the stent implantation, in which the stent was modeled by SMA and the vessel was defined as Mooney-Rivlin material.

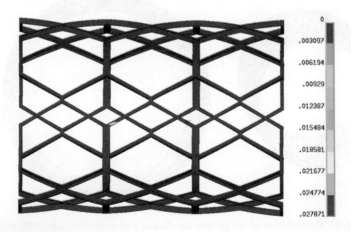

FIGURE 16.11   Plastic strains of the stent.

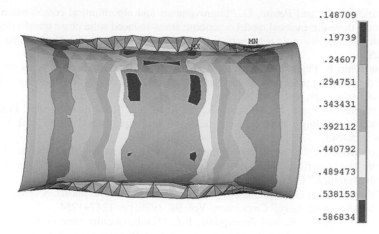

.148709
.19739
.24607
.294751
.343431
.392112
.440792
.489473
.538153
.586834

FIGURE 16.12   von Mises stresses of the plaque (MPa).

The computational results show that after the stent is inserted within the vessel and contacts the plaque, both the vessel and the plaque get enlarged to allow more blood flow, which indicates that the built finite element model simulates angioplasty with vascular stenting.

The computational results also indicate that after the stent contacts the vessel, plastic strains exist in the stent; this demonstrates that the stent undergoes phase transformation. The mechanical state of the stent is associated with the thickness of the plaque. With a larger thickness of the plaque, the plaque and vessel undergo larger deformation with the stretch of the stent within the vessel, which consequently causes greater stresses and plastic strains in the stent.

Two SMA material models are available in ANSYS. This study does not involve temperature change. Thus, the SMA model for superelasticity (TB, SMA, , , , SUP) was applied to directly simulate the stent. This simulation can also be accomplished by SMA with the shape memory effect (TB, SMA, , , , MEFF) and with constant temperature specified in the model.

16.2.4 SUMMARY

Angioplasty with vascular stenting was simulated in ANSYS using the SMA material model. The computational results indicate that the built finite element model can be applied to study the stent implantation.

REFERENCES

1. Auricchio, F., Taylor, R. L., and Lubliner, J., "Shape-memory alloys: Macromodeling and numerical simulations of the superelastic behavior." *Computational Methods in Applied Mechanical Engineering*, Vol. 146, 1997, pp. 281–312.
2. Souza, A. C., Mamiya, E. N., and Zouain, N., "Three-dimensional model for solids undergoing stress-induced phase transformations." *European Journal of Mechanics-A/ Solids*, Vol. 17, 1998, pp. 789–806.

3. Auricchio, F., and Petrini, L., "Improvements and algorithmical considerations on a recent three-dimensional model describing stress-induced solid phase transformations." *International Journal for Numerical Methods in Engineering*, Vol. 55, 2005, pp. 1255–1284.

4. Auricchio, F., Fugazza, D., and DesRoches, R., "Numerical and experimental evaluation of the damping properties of shape-memory alloys." *Journal of Engineering Materials and Technology*, Vol. 128, 2006, pp. 312–319.

5. Auricchio, F., Conti, M., Morganti, S., and Reali, A., "Shape memory alloy: From constitutive modeling to finite element analysis of stent deployment." *CMES*, Vol. 57, 2010, pp. 225–243.

6. Lloyd-Jones, D. et al., "Heart disease and stroke statistics—2009 update: A report from the American Heart Association Statistics Committee and Stroke Statistics Subcommittee." *Circulation*, Vol. 119, 2009, pp. 1–161.

7. Hoffmann, R. et al., "Patterns and mechanisms of in-stent restenosis. A serial intravascular ultrasound study." *Circulation*, Vol. 94, 1996, pp. 1247–1254.

8. Lally, C., Dolan, F., and Pendergrast, P.J., "Cardiovascular stent design and vessel stresses: A finite element analysis." *Journal of Biomechanics*, Vol. 38, 2005, pp. 1574–1581.

17 Wear Model of Liner in Hip Replacement

Chapter 17 focuses on wear simulation of the liner in hip replacement. Wear modeling in ANSYS is introduced in Section 17.1, followed by wear simulation of the liner in Section 17.2.

17.1 WEAR SIMULATION

17.1.1 ARCHARD WEAR MODEL

The Archard wear equation is as follows [1]:

$$\dot{w} = \frac{K}{H}P^m v^n \tag{17.1}$$

where $\dot{w}$ is the rate of wear $= w/L$, L is the sliding distance, K is the wear coefficient, H is material hardness, P is the contact pressure, v is the sliding velocity, m is the pressure exponent, and n is the velocity exponent.

In ANSYS190, it is defined by

```
TB, WEAR, 1, , ,ARCD
TBDATA, 1, K, H, m, n, C5
TBDATA, 6, nx, ny, nz
```

In this definition, $C5$ is an option flag for contact pressure calculation, and nx, ny, and nz indicate the wear directions.

17.1.2 IMPROVING MESH QUALITY DURING WEAR

During wear, the surface changes to simulate material loss. With increasing wear, the element quality at the wear surface worsens. The serious element distortion may cause termination of analysis. Mesh nonlinear adaptivity, a powerful tool for remeshing during solution, is a good technique to improve mesh quality during wear. The commands have the following format:

```
! morph the mesh after nn% is lost in wear
NLADAPTIVE, CONTWEAREL, ADD, CONTACT, WEAR, NN/100
NLADAPTIVE, CONTWEAREL, ON, , , NSTP, T1, T2
! criterion defined for component contwearel becomes active from
time t1 to t2
!and check once every nstp substep.
```

17.2 SIMULATING WEAR OF LINER IN HIP REPLACEMENT

As Americans age, more and more individuals need joint replacements, such as a total hip replacement (Figure 17.1). The Agency for Healthcare Research and Quality estimated that about 250,000 total hip replacements are performed in the United States each year [2]. This significant demand for hip replacements inspired the development and production of hip-resurfacing devices, as well as more research into the hip replacement procedure. In the past, research primarily focused on wear tests to predict the mechanical wear of the implant devices. In addition, some finite element models were developed to simulate the wear of the hip replacement [3–6]. In this study, the wear of the hip replacement in ANSYS190 was simulated.

17.2.1 FINITE ELEMENT METHOD

17.2.1.1 Geometry and Mesh

The hip implant is comprised of the acetabular shell, polyethylene liner, femoral head, neck, and stem (Figure 17.2). The whole model, which was meshed by SOLID187 in the workbench, contains 9,631 elements and 14,957 nodes. In the model, the femoral head, neck, and stem are glued; they share the common nodes at the interface.

17.2.1.2 Material Properties

The liner and metal were assumed to be linear elastic. The liner was defined with a Young's modulus of 1,200 MPa and a Poisson's ratio of 0.4, while the metal was specified as having a Young's modulus of 58,000 MPa and a Poisson's ratio of 0.3.

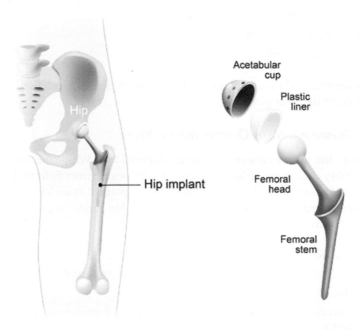

FIGURE 17.1 Schematic of a hip implant. (Designua © 123RF.com.)

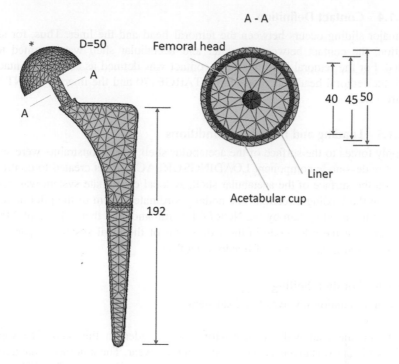

FIGURE 17.2 Finite element model of the hip implant (all dimensions in millimeters).

17.2.1.3 Wear Model

The Archard wear model was applied to define the wear of the liner. The sliding distance on the liner surface is πR. The wear coefficient is assumed as $3e - 9$. To estimate the wear of 1e6 cycles, the input data wear coefficient was calculated by

$$K = 3e - 9*1e6*3.1416*25 \tag{17.2}$$

```
kUW = 3e-9*1e6*3.1416*25
TB, WEAR, 5,,,ARCD              ! Mat5 for wear of contact element on
                                  UHMWPE
TBFIELD, TIME, 0
TBDATA,1,0,1,1,0,0              ! no wear for load step#1;
TBFIELD, TIME, 1
TBDATA,1,0,1,1,0,0
TBFIELD, TIME, 1.01            ! start wear in load step #2
TBDATA, 1, kUW, 1, 1, 0, 0
TBFIELD, TIME, 2
TBDATA, 1, kUW, 1, 1, 0, 0
```

17.2.1.4 Contact Definition

The major sliding occurs between the femoral head and the liner. Thus, for simplification, the contact between the liner and acetabular shell was assumed to be bonded. For the femoral head, the liner contact was defined as standard contact, in which the femoral head was modeled by TARGE170 and the liner by CONTA174 (Figure 17.3).

17.2.1.5 Loading and Boundary Conditions

To apply forces to the surface of the acetabular shell, MPC constraints were set up. A pilot node—node component LOADINGSURFACE—was created to control the nodes on the surface of the acetabular shell. A local coordinate system was built to align with the loading direction. The nodal coordinate system of the pilot node was rotated to the local system by the NORTAT command, and then a force of 2,000 N was loaded on the pilot node in the *x*-direction of the local system. A part of the stem was fixed in all degrees of freedom (DOFs).

17.2.1.6 Solution Setting

The wear simulation involves two load steps:

1. Ramp the load without wear—force was loaded in this step. The wear coefficient was given as zero, to simulate no wear. The nonlinear adaptivity criterion was defined on the assumption that morph occurs after 50% was lost in wear.
2. Simulate wear—a very small-time step (1e-6) was selected as a minimum.

(a)   (b)

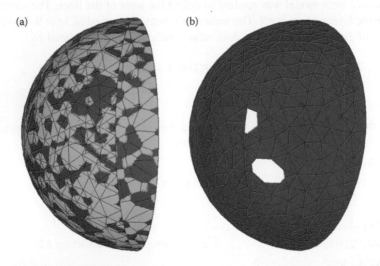

FIGURE 17.3 Contacts in the hip implant. (a) Standard contact between the femoral head (in green) and the liner (in red); (b) bonded contact between the liner and acetabular shell to share common nodes.

```
/SOLU
! load step 1: ramp to the applied force (no wear)
NLGEOM, ON
ALLSEL, ALL
OUTRES, ALL, ALL
! define nonlinear adaptivity criterion
NLAD, CONWEAREL, ADD, CONTACT, WEAR, 0.50     ! morph after 50% is
                                                lost in wear
NLAD, CONWEAREL, ON, ALL, ALL, 1, ,2
NLAD, CONWEAREL, LIST, ALL, ALL
TIME,1
DELTIM,0.1,1E-4,1
SOLVE

! load Step 2: start the wear
TIME, 2
DELTIM, 0.01, 1E-6, 0.02
NLHIST, PAIR, WV, CONT, WEAR, 5
SOLVE
FINISH
```

17.2.2   Results

Figure 17.4 illustrates the contact stresses of the liner during wear. The contact stress starts at a low value, increases to a peak value, and then drops a little.

Using the following command, the wear contour of the linear surface is plotted as shown in Figure 17.5:

```
ESEL, S, TYPE, ,5
NSLE, S
ESLN, S
ESEL, R, MAT, ,5
ETABLE, WEAR, NMISC, 176
PLETAB, WEAR
```

The wear primarily occurs in the center, which is consistent with the stress distribution.

17.2.3   Discussion

A three-dimensional (3D) finite element model was created to simulate the wear of the liner. The computational results show that the contact pressures of the liner vary with

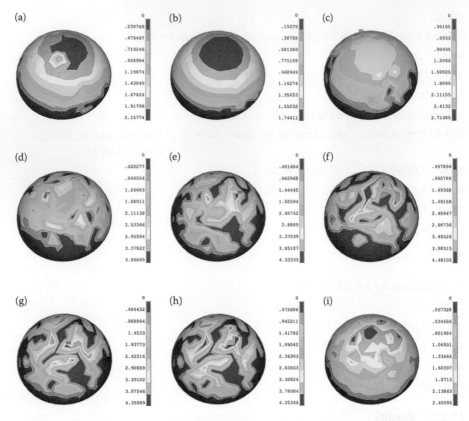

FIGURE 17.4 Contact pressures at different time steps (MPa) (T is the total time). (a) At T/9; (b) at 2T/9; (c) at 3T/9; (d) at 4T/9; (e) at 5T/9; (f) at 6T/9; (g) at 7T/9; (h) at 8T/9; (i) at T.

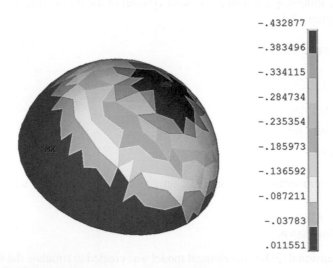

FIGURE 17.5 Wear contour of the liner at the final step (mm).

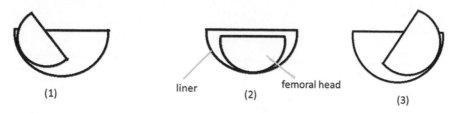

(1)

liner

(2)

femoral head

(3)

FIGURE 17.6 Schematic to compute the wear during one cycle (sliding distance L).

$$w = \dot{w}_1 \cdot \frac{L}{3} + \dot{w}_2 \cdot \frac{L}{3} + \dot{w}_3 \cdot \frac{L}{3}$$

time. This is a standard wear procedure. When wear starts, the contact surface becomes rougher and the contact pressures increase. Furthermore, the higher contact pressures cause more material loss, which make the contact surface smoother. That is why the contact pressures drop a little after reaching a peak value. This procedure repeats during the whole wear stage.

The big drawback of this study is that the wear of the liner was simulated using one position of the liner. The liner slides on the femoral head and the contact pressure between the liner and the femoral head vary during the movement of the liner. Therefore, to obtain more accurate results, a different approach was proposed: to split one cycle of the liner motion into several parts, in which each part is represented by a different contact position of the liner relative to the femoral head (Figure 17.6). Thus, the whole wear of the liner at one cycle is the superposition of wear results of the liner at various positions.

17.2.4 SUMMARY

The wear of the liner of a hip implant was simulated in ANSYS190. The computational results match the general wear procedure. The developed finite element model can be improved using the superposition method.

REFERENCES

1. ANSYS19.0 Help documentation in the help page of product ANSYS190.
2. American Academy of Orthopaedic Surgeons, *Total Hip Replacement*. Accessed on December 3, 2012, from http://orthoinfo.aaos.org/topic.cfm?topic=a00377
3. Liu, F., Leslie, I., Williams, S., Fisher, J., and Jin, Z., "Development of computational wear simulation of metal-on-metal hip resurfacing replacements." *Journal of Biomechanics*, Vol. 41, 2008, pp. 686–694.
4. Meng, H. C., and Ludema, K., "Wear models and predictive equations: Their form and content." *Wear*, Vol. 181–183, 1995, pp. 443–457.
5. Archard, J. F., "Contact and rubbing of flat surfaces." *Journal of Applied Physics*, Vol. 24, 1953, pp. 981–988.
6. Ronda, J., and Wojnarowski, P., "Analysis of wear of polyethylene hip joint cup related to its positioning in patient's body." *Acta of Bioengineering and Biomechanics*, Vol. 15, 2013, pp. 77–86.

18 Fatigue Analysis of a Mini Dental Implant (MDI)

Different from the traditional fatigue analysis in mechanical design, a new finite element technology based on crack growth has been developed for fatigue analysis, which is introduced in Section 18.1; its application on dental implant appears in Section 18.2.

18.1 SMART CRACK-GROWTH TECHNOLOGY

Separating, Morphing, Adaptive, and Remeshing Technology (SMART) was developed in ANSYS190 to simulate fatigue crack growth [1]. Unlike the eXtended Finite Element Method (XFEM), SMART always generates new elements and nodes at the crack front after remeshing (Figure 18.1). In SMART, cracks grow based on Paris' Law. Generally, SMART requires users to define the fracture parameter calculation set and fatigue crack-growth calculation set.

a. *Defining the Fracture-Parameter Calculation Set*
   The following commands are used to define the fracture parameter calculation:

```
CINT, NEW, SETNUMBER              ! define the crack calculation set

CINT, TYPE, Fracture Parameter    ! define the type of fracture
                                    parameter

CINT, CTNC, PAR1                  ! specify crack front node
                                    component

CINT, NCON, NUM_CONTOURS          ! specify the number of contours for
                                    fracture-parameter calculation
```

b. *Specifying Fatigue Crack-Growth Calculation Set*
   Fatigue crack-growth calculation requires the following input:

```
CGROW, NEW, SETNUMBER             ! define a set number for crack-growth
                                    calculation.

CGROW, CID, ID                    ! specify the crack calculation ID

CGROW, METHOD, SMART, REME        ! define the crack-growth method SMART

CGROW, FCG, METH, LC or CBC       ! specify either the life cycle (LC)
                                    or cycle by cycle

                                  ! method for fatigue crack growth

TB, CGCR, MAT_ID,,, Option        ! define the fatigue parameters
```

177

(a)

(b)

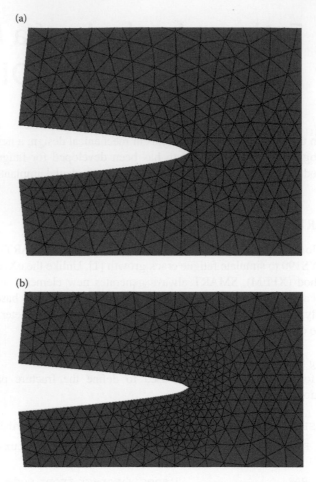

FIGURE 18.1 Schematic of SMART. (a) Before crack growth; (b) after crack growth. Remeshing occurs around the crack front.

18.2 STUDY OF FATIGUE LIFE OF A MDI

Implantology has improved the therapy of edentulous patients. A total of 96% of oral rehabilitation has been proven successful using two to four implants to support the complete prosthesis. The standard implants have a diameter ranging from 2.75–3.30 mm, which is not always suitable for surgery due to the atrophy of an edentulous jaw, especially in the lower jaw area [2]. Also, the large size of the implant requires an aggressive surgical procedure, including cutting the gingiva and lifting the chop for bone preparation. This procedure causes many issues, such as tissue healing, the recovery of vascular function, and implant osseointegration [3]. Mini dental implants (MDIs), with a diameter ranging from 1.8–2.4 mm, have an advantage in this case [4]. The surgery of the MDI can be completed in one visit with minimally invasive surgical procedures [5]. However, it was found that cracks may be initiated on the implant's surface during installation of the MDI microcracks; their growth can lead to the fatigue failure of the MDI under cycle loading [6]. To investigate this, a finite element model of the MDI was created, and its crack growth was studied in ANSYS190.

18.2.1 FINITE ELEMENT MODEL

18.2.1.1 Geometry and Mesh

A three-dimensional (3D) geometry of the MDI was created in SpaceClaim and meshed with SOLID187. Then, it was transferred to the Workbench; an elliptic crack was added to represent the initial crack (Figure 18.2). The whole model comprises 82,359 elements and 114,598 nodes.

18.2.1.2 Material Properties

MDI was assumed to be linear elastic, with a Young's modulus of 200 GPa and a Poisson's ratio of 0.33. Paris' s Law with $C = 2.29e\text{-}10$ and $M = 2$ govern the crack growth as follows:

```
! Fatigue Crack-Growth Law Specification
TB, CGCR, 2, , , PARIS
TBDATA, 1, C, M
```

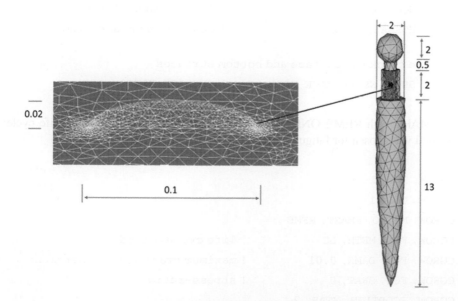

FIGURE 18.2 Finite element model of MDI (all dimensions in millimeters)

18.2.1.3   Loading and Boundary Conditions

Some studies found that with edentulous patients, the magnitude of force varied from 50 N to 210 N [7]. Therefore, this study selected the average value of 100 N as the loading amplitude applied on the top surface of the MDI in the y-direction. The bottom part of the MDI was constrained in all degrees of freedom (DOFs).

18.2.1.4   Setting up Fracture Calculation

The stress intensity factor (SIFS) was selected for the fracture parameter. In the fracture parameter calculation, the crack front was specified by CINT, CTNC in the form of the crack tip node component. An assistant node on the open side of the crack was added to help define the crack extension direction. Unlike the normal fracture parameter calculation, the top and bottom surfaces of the crack were required to be defined by CINT, SURF in SMART. The normal direction of the crack was aligned with the y-axis of the local coordinate system 13:

```
*SET,_IASSISTNODE,115121
*SET,_SIFS,1
CINT, NEW, 1                          ! define crack id for
                                      Semi-Elliptical Crack

CINT, TYPE, SIFS                      ! output quantity for
                                      Semi-Elliptical Crack

CINT, CTNC, NS_SECRACK_FRONT,
_IASSISTNODE                          ! define crack tip node component
CINT, NCON, 6                         ! define number of contours
CINT, NORMAL, 13, 2                   ! define crack plane normal


! define crack top surface and bottom surfaces
CINT, SURF, NS_SECRACK_TOPFACE, NS_SECRACK_BOTTOMFACEx
```

SMART with REME ONLY was selected for crack growth, and the life cycle method was chosen for fatigue analysis:

```
CGROW, NEW, 1
CGROW, CID, 1
CGROW, METHO, SMART, REME
CGROW, FCG, METH, LC                  ! life cycle method
CGROW, FCG, DAMX, 0.01                ! maximum crack growth increment
CGROW, FCG, SRAT, 0                   ! stress-ratio
CGROW, FCOPTION, MTAB, 2
```

18.2.2 RESULTS

The computation was completed in five steps, in which four remeshings occurred as the crack grew. The crack front at steps 1, 3, and 5 is plotted in Figure 18.3. Figure 18.4 shows the stress intensity factor K as path independent.

ANSYS obtained data of deltK, deltA, and deltN in the postprocess by the command *Get. The obtained deltA and deltN were added onto the previous crack length and number of cycles to get the outcome of the current crack length and number of cycles.

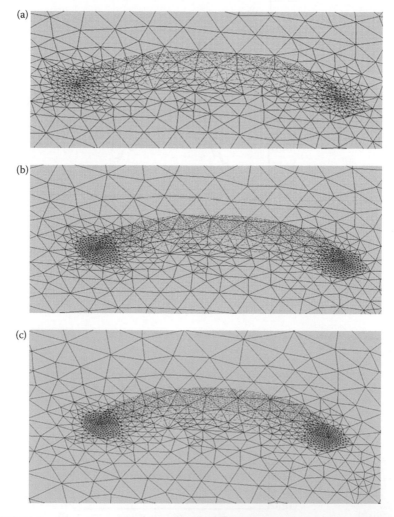

FIGURE 18.3   Crack front of MDI model at different steps. (a) Step 1 (dela = 0.00058 mm); (b) step 3 (dela = 0.00179 mm); (c) step 5 (delta = 0.00311 mm).

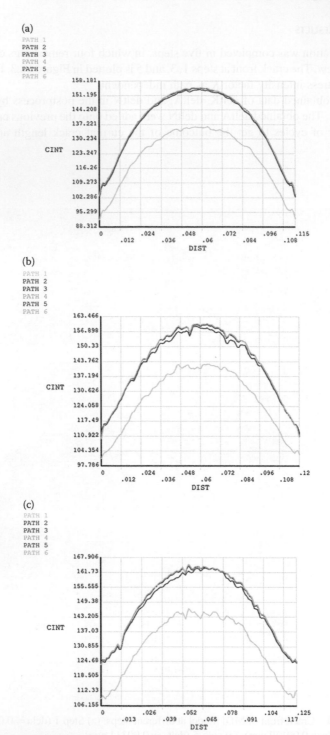

FIGURE 18.4   *K1* of the crack front at different steps. (a) Step 1; (b) step 3; (c) step 5.

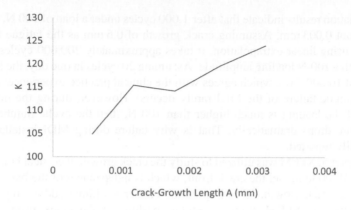

FIGURE 18.5 *A–K* curve of the crack.

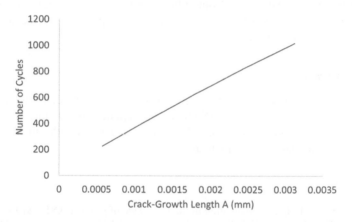

FIGURE 18.6 *A–N* curve of the crack.

```
*GET, FVAL, CINT, CRKID, CTIP, NDNUM, CONTOUR, 1, DTYPE, DLTN
DTN(INODE, ISTEP) = DTN(INODE, ISTEP-1) + FVAL
*GET, FVAL, CINT, CRKID, CTIP, NDNUM, CONTOUR, 1, DTYPE, DLTA
DTA(INODE, ISTEP) = DTA(INODE, ISTEP-1) + FVAL
*GET, FVAL, CINT, CRKID, CTIP, NDNUM, CONTOUR, 1, DTYPE, DLTK
DTK(INODE, ISTEP) = FVAL
```

Therefore, curves *A–K* and *A–N* in the entire crack growth are plotted in Figures 18.5 and 18.6, respectively, which help to determine the fatigue life of the MDI.

18.2.3 Discussion

The crack growth in the MDI was simulated in ANSYS190 using SMART technology. The stress intensity factor *K1* at the crack front stayed path-independent.

The simulation results indicate that after 1,000 cycles under a load of 100 N, the crack grows about 0.003 mm. Assuming crack growth of 0.6 mm as the fatigue failure of the MDI, using linear extrapolation, it takes approximately 200,000 cycles to reach failure with a 100-N loading amplitude. Assuming 20 cycles in one day, the MDI fails after about 10,000 days, which agrees with the clinical practice experiences (namely, that the fatigue failure of the MDI rarely occurs). However, during the installation process, if the loading is much higher than 100 N, then the cyclic number for the MDI failure drops dramatically. That is why failure during MDI installation was occasionally reported.

In Chapter 5, XFEM was utilized to study the crack growth of cortical bone. XFEM uses MESH200 to model the crack front, which is independent of the base meshing. Also, in the crack growth, XFEM splits the elements without adding any new elements. However, XFEM requires regular meshing at the crack front, which is not applicable in most practical problems. Although when compared to XFEM, SMART needs to create the crack front in the geometry and perform remeshing while the cracks grow, it still works for all kinds of meshing, especially free meshing, which exists in most practical problems.

18.2.4 SUMMARY

Fatigue and wear represent two significant concerns for some metal implants. In this chapter, SMART technology in ANSYS190 was applied to simulate crack growth of the MDI. Compared to XFEM, SMART does not require regular meshing around the crack front and has a wider application in industry.

REFERENCES

1. ANSYS 19.0 Help Documentation in the help page of product ANSYS190.
2. Meijer, H. J., Starmans, F. J., Bosman, F., and Steen, W. H., "A comparison of three finite element models of an edentulous mandible provided with implants." *Journal of Oral Rehabilitation*, Vol. 20, 1993, pp. 147–157.
3. Zarb, G. A., and Schmitt, A., "Osseointegration for elderly patients: The Toronto study." *Journal of Prosthetic Dentistry*, Vol. 72, 1994, pp. 559–568.
4. Shatkin, T. E., Shatkin, S., Oppenheimer, B. D., and Oppenheimer, A. J., "Mini dental implants for long-term fixed and removable prosthetics: A retrospective analysis of 2514 implants placed over a five-year period." *Compendium of Continuing Education in Dentistry*, Vol. 28, 2007, pp. 36–43.
5. Gibney, J. W., "Minimally invasive implant surgery." *Journal of Oral Implantology*, Vol. 27, 2001, pp. 73–76.
6. Grbović, A. M., Rašuo, B. P., Vidanović, N. D., and Perić, M. M., "Simulation of crack propagation in titanium mini dental implants (MDI)." *FME Transactions*, Vol. 39, 2011, pp. 165–170.
7. Takayama, Y., Yamada, T., Araki, O., Seki, T., and Kawasaki, T., "The dynamic behavior of a lower complete during unilateral loads: Analysis using the finite element method." *Journal of Oral Rehabilitation*, Vol. 28, 2001, pp. 1064–1074.

Part V

Retrospective

The previous four parts presented the finite element modeling of bone, soft tissues, joints, and implants. Based on this content, Part V summarized the principle for modeling biology, including relative stiffness, time factor, macroscale and microscale, and contact behavior. In addition, it discussed mesh sensitivity, units, workbench, and ANSYS versions.

19 Retrospective

19.1 PRINCIPLES FOR MODELING BIOLOGY

Unlike materials in industry, biomaterials are very complicated. As discussed in Part I, bone is nonlinear, anisotropic, and nonhomogeneous. Soft tissues are porous media, in which the solid phase is nonlinear and viscoelastic. The contacts at the joints may have different contact behaviors, such as standard contact and always bonded. Therefore, to build an accurate finite element model that close to the true biology, a number of factors should be considered.

1. *Relative Stiffness*: One material may be modeled differently in each model, depending on its relative stiffness. In the ankle replacement problem in Chapter 15, the talar component made of titanium alloys is approximately 1,000 times stiffer than the bone. In this case, material properties of the bone, implemented by the TBFIELD command, vary across the whole knee. However, in Section 13.2 of Chapter 13, which focuses on the contact in the whole knee, the bone is more than 100 times stiffer than the cartilage and meniscus. Here, the bone is regarded as rigid and its motion is controlled by a pilot node. Thus, the bone is defined as linear elastic with one material ID.

2. *Time*: Time is always associated with the viscoelastic behavior of materials. The soft tissues are porous media. Section 11.3 of Chapter 11 describes the poroelastic creep response of an intervertebral disc (IVD) under compression. The disc's upper surface moves down gradually; after 5 days, it approaches a constant value. The IVD is modeled with coupled pore-pressure thermal (CPT) elements to simulate the internal friction between the fluid and solid phases. Section 9.1 of Chapter 9 examines the mechanical behavior of the IVD after the nucleus of the IVD swells. Because the permeability of the IVD is extremely low, the soft tissue of the IVD is modeled as solid only to study its instant response.

3. *Macroscale/Microscale*: Generally, bone is considered as anisotropic and nonhomogeneous. Thus, it does not meet the basic requirements of fracture calculation; as a result, fracture calculation cannot work with bone on the macroscale. However, on the microscale, the cortical bone is assumed to be isotropic and homogeneous; XFEM is applied to study its crack growth in Chapter 5.

4. *Contact*: Contact exists everywhere, especially in joints. Contact behavior is generally classified into three types, as described next.
   - *Standard*: One example of standard contact is cartilages—meniscus contact. The contact is modeled by CONTAC174/175 with Keyopt(12) = 0.

- *Always bonded*: One example of always-bonded contact is contact between femoral cartilages and the femur. The contact is defined by CONTAC174/175 with keyopt(12) = 5.
- *MPC*: One example is rigid bone controlled by a pilot node. The contact is specified by CONTAC174/175 with keyopt(2) = 2.

  Another example is a force-distributed constraint applied to the surfaces of the artery to allow radial arterial expansion. The contact is modeled as CONTAC174/175 with keyopt(2) = 2 and keyopt(4) = 1.

19.2 MESHING SENSITIVITY

The meshing sensitivity of the finite element models in the book was verified by a mesh refinement. After increasing the mesh density, the maximum von Mises (vM) stresses of the models gradually reached constants. Thus, the convergence of the finite element models was validated.

19.3 UNITS

The size of human body parts is measured in millimeters. Thus, units (mm, N, MPa, s) are selected for units of length, force, pressure, and time in the models of this book. For example, the permeability of soft tissues is around $1e-14 \, m^4/N/s$. In the current units, it becomes $1e-2 \, mm^4/N/s$. Therefore, the value 0.01 is input into the TBDATA for the TB, PM,,,,PERM command.

19.4 WORKBENCH

All models in the book are implemented in ANSYS Parametric Design Language (APDL), which easily explains the modeling process. However, ANSYS Workbench is also very popular, especially for the younger generation. ANSYS Help Documentation makes it easy to find the corresponding operation in Workbench for APDL commands and to complete the models of the book in Workbench. One alternative to this approach is to copy APDL commands directly into Workbench to create the model.

19.5 ANSYS VERSIONS

Over the last three years, ANSYS Mechanical APDL (MAPDL) has released some major features, such as the eXtended Finite Element Method (XFEM), Separating, Morphing, Adaptive, and Remeshing Technology (SMART), wear simulation, mesh-independent fiber enhancement, and multidimensional interpolation. These features are applied to some of the examples in this book. Therefore, these examples should be repeated in ANSYS190 and later versions.

Appendix 1: Input File of the Multidimensional Interpolation in Section 3.2.2

```
/COM, IT CONTAINS TWO PARTS:
/COM, THE FIRST ONE IS FOR LMUL-RBAS_NNEI ALGORITHM
/COM, THE SECOND ONE IS FOR MIXED ALGORITHM
/COM
/COM, INPUT FILE FOR LMUL_RBAS_NNEI ALGORITHM
/COM, download ankle.db from
/COM, www.feabea.net/models/ankle.db
/COM, the model can be repeated in ANSYS190 and later versions
/PREP7
RESUME, 'ankle', 'db'
TB, ELAS, 6
TBFIELD, XCOR, 191.75
TBFIELD, YCOR, -22.951
TBFIELD, ZCOR, 135
TBDATA, 1, 111 , 0.30

TBFIELD, XCOR, 185.36
TBFIELD, YCOR, -33.449
TBFIELD, ZCOR, 135
TBDATA, 1, 134 , 0.30

TBFIELD, XCOR, 185.36
TBFIELD, YCOR, -33.449
TBFIELD, ZCOR, 145
TBDATA, 1, 120 , 0.30

TBFIELD, XCOR, 177.85
TBFIELD, YCOR, -44.736
```

```
TBFIELD, ZCOR, 135
TBDATA, 1, 83 , 0.30


TBFIELD, XCOR, 177.85
TBFIELD, YCOR, -44.736
TBFIELD, ZCOR, 145
TBDATA, 1, 132 , 0.30


TBFIELD, XCOR, 177.05
TBFIELD, YCOR, -18.441
TBFIELD, ZCOR, 135
TBDATA, 1, 469 , 0.30


TBFIELD, XCOR, 177.05
TBFIELD, YCOR, -18.441
TBFIELD, ZCOR, 145
TBDATA, 1, 604 , 0.30


TBFIELD, XCOR, 171.15
TBFIELD, YCOR, -30.939
TBFIELD, ZCOR, 135
TBDATA, 1, 191 , 0.30


TBFIELD, XCOR, 171.15
TBFIELD, YCOR, -30.939
TBFIELD, ZCOR, 145
TBDATA, 1, 68 , 0.30


TBFIELD, XCOR, 166.77
TBFIELD, YCOR, -42.402
TBFIELD, ZCOR, 135
TBDATA, 1, 160 , 0.30


TBFIELD, XCOR, 166.77
TBFIELD, YCOR, -42.402
TBFIELD, ZCOR, 145
TBDATA, 1, 125 , 0.30


TBFIELD, XCOR, 160.38
TBFIELD, YCOR, -15.732
```

```
TBFIELD, ZCOR, 135
TBDATA, 1, 602 , 0.30

TBFIELD, XCOR, 160.38
TBFIELD, YCOR, -15.732
TBFIELD, ZCOR, 145
TBDATA, 1, 357 , 0.30

TBFIELD, XCOR, 155.90
TBFIELD, YCOR, -29.259
TBFIELD, ZCOR, 135
TBDATA, 1, 192 , 0.30

TBFIELD, XCOR, 155.90
TBFIELD, YCOR, -29.259
TBFIELD, ZCOR, 145
TBDATA, 1, 393 , 0.30

TBFIELD, XCOR, 151.25
TBFIELD, YCOR, -40.088
TBFIELD, ZCOR, 135
TBDATA, 1, 90 , 0.30

TBFIELD, XCOR, 151.25
TBFIELD, YCOR, -40.088
TBFIELD, ZCOR, 145
TBDATA, 1, 158 , 0.30

TBFIELD, XCOR, 149.06
TBFIELD, YCOR, -15.388
TBFIELD, ZCOR, 135
TBDATA, 1, 403 , 0.30

TBFIELD, XCOR, 149.06
TBFIELD, YCOR, -15.388
TBFIELD, ZCOR, 145
TBDATA, 1, 659 , 0.30

TBFIELD, XCOR, 145.38
TBFIELD, YCOR, -25.952
```

```
TBFIELD, ZCOR, 135
TBDATA, 1, 407 , 0.30


TBFIELD, XCOR, 145.38
TBFIELD, YCOR, -25.952
TBFIELD, ZCOR, 145
TBDATA, 1, 365 , 0.30


TBFIELD, XCOR, 140.79
TBFIELD, YCOR, -35.642
TBFIELD, ZCOR, 135
TBDATA, 1, 351 , 0.30


TBFIELD, XCOR, 140.79
TBFIELD, YCOR, -35.642
TBFIELD, ZCOR, 145
TBDATA, 1, 222 , 0.30


! TBIN, ALGO, LMUL
! TBIN, ALGO, NNEI
TBIN, ALGO, RBAS
ESEL, S, MAT, , 5
EMODIF, ALL, MAT, 6
NSLE, S
*GET, NUM_N, NODE, 0, COUNT ! GET NUMBER OF NODES
*GET, N_MIN, NODE, 0, NUM, MIN ! GET MIN NODE NUMBER
*DO, I, 1, NUM_N, 1      ! OUTPUT TO ASCII BY LOOPING OVER NODES
        CURR_N = N_MIN
        DZZ = NZ(CURR_N) - 150.45
        D, CURR_N, UZ, DZZ
        D, CURR_N, UX, 0
        D, CURR_N, UY, 0
        *GET, N_MIN, NODE, CURR_N, NXTH
*ENDDO
FINISH
/SOLU
ERESX, NO
```

```
NSUBST, 1, 1, 1
SOLV
FINISH

*************************************************************

/COM, INPUT FILE FOR MODIFIED LMUL MODEL
/PREP7
RESUME, 'ankle', 'db'
ESEL, S, MAT, , 5
*GET, NUM_E, ELEM, 0, COUNT ! GET NUMBER OF ELEMENTS
*GET, E_MIN, ELEM, 0, NUM, MIN ! GET MIN ELEMENT NUMBER

*DO, I, 1, NUM_E
CURR = E_MIN
XX = CENTRX(CURR)
YY = CENTRY(CURR)
N1X = 191.75
N1Y = -22.951
N2X = 149.06
N2Y = -15.388
N3X = 140.79
N3Y = -35.642
N4X = 177.85
N4Y = -44.736
S = 0
 *DO, J, 1, 4
  *IF, J, EQ, 1, THEN
          X1 = XX
          Y1 = YY
          X2 = N1X
          Y2 = N1Y
          X3 = N4X
          Y3 = N4Y
 *ELSEIF, J, EQ, 2
          X1 = XX
          Y1 = YY
          X2 = N1X
```

```
            Y2 = N1Y
            X3 = N2X
            Y3 = N2Y
*ELSEIF, J, EQ, 3
            X1 = XX
            Y1 = YY
            X2 = N2X
            Y2 = N2Y
            X3 = N3X
            Y3 = N3Y
*ELSE
            X1 = XX
            Y1 = YY
            X2 = N3X
            Y2 = N3Y
            X3 = N4X
            Y3 = N4Y
*ENDIF
SS0 = X1*Y2 + X2*Y3 + X3*Y1
SS1 = Y1*X2 + Y2*X3 + Y3*X1
SS = ABS((SS0 - SS1)/2)
S = S + SS
*ENDDO
DELS = ABS(S - 930.474)
*IF, DELS, GT, 10, THEN
        EMODIF, CURR, MAT, 6
*ENDIF
*GET, E_MIN, ELEM, CURR, NXTH
*ENDDO

/PREP7
TBDE, ALL, 5
TB, ELAS, 5
TBFIELD, XCOR, 191.75
TBFIELD, YCOR, -22.951
TBFIELD, ZCOR, 135
TBDATA, 1, 111 , 0.30
```

```
TBFIELD, XCOR, 185.36
TBFIELD, YCOR, -33.449
TBFIELD, ZCOR, 135
TBDATA, 1, 134 , 0.30


TBFIELD, XCOR, 185.36
TBFIELD, YCOR, -33.449
TBFIELD, ZCOR, 145
TBDATA, 1, 120 , 0.30


TBFIELD, XCOR, 177.85
TBFIELD, YCOR, -44.736
TBFIELD, ZCOR, 135
TBDATA, 1, 83 , 0.30


TBFIELD, XCOR, 177.85
TBFIELD, YCOR, -44.736
TBFIELD, ZCOR, 145
TBDATA, 1, 132 , 0.30


TBFIELD, XCOR, 177.05
TBFIELD, YCOR, -18.441
TBFIELD, ZCOR, 135
TBDATA, 1, 469 , 0.30


TBFIELD, XCOR, 177.05
TBFIELD, YCOR, -18.441
TBFIELD, ZCOR, 145
TBDATA, 1, 604 , 0.30


TBFIELD, XCOR, 171.15
TBFIELD, YCOR, -30.939
TBFIELD, ZCOR, 135
TBDATA, 1, 191 , 0.30


TBFIELD, XCOR, 171.15
TBFIELD, YCOR, -30.939
TBFIELD, ZCOR, 145
TBDATA, 1, 68 , 0.30
```

```
TBFIELD, XCOR, 166.77
TBFIELD, YCOR, -42.402
TBFIELD, ZCOR, 135
TBDATA, 1, 160 , 0.30


TBFIELD, XCOR, 166.77
TBFIELD, YCOR, -42.402
TBFIELD, ZCOR, 145
TBDATA, 1, 125 , 0.30


TBFIELD, XCOR, 160.38
TBFIELD, YCOR, -15.732
TBFIELD, ZCOR, 135
TBDATA, 1, 602 , 0.30


TBFIELD, XCOR, 160.38
TBFIELD, YCOR, -15.732
TBFIELD, ZCOR, 145
TBDATA, 1, 357 , 0.30


TBFIELD, XCOR, 155.90
TBFIELD, YCOR, -29.259
TBFIELD, ZCOR, 135
TBDATA, 1, 192 , 0.30


TBFIELD, XCOR, 155.90
TBFIELD, YCOR, -29.259
TBFIELD, ZCOR, 145
TBDATA, 1, 393 , 0.30


TBFIELD, XCOR, 151.25
TBFIELD, YCOR, -40.088
TBFIELD, ZCOR, 135
TBDATA, 1, 90 , 0.30


TBFIELD, XCOR, 151.25
TBFIELD, YCOR, -40.088
```

```
TBFIELD, ZCOR, 145
TBDATA, 1, 158 , 0.30


TBFIELD, XCOR, 149.06
TBFIELD, YCOR, -15.388
TBFIELD, ZCOR, 135
TBDATA, 1, 403 , 0.30


TBFIELD, XCOR, 149.06
TBFIELD, YCOR, -15.388
TBFIELD, ZCOR, 145
TBDATA, 1, 659 , 0.30


TBFIELD, XCOR, 145.38
TBFIELD, YCOR, -25.952
TBFIELD, ZCOR, 135
TBDATA, 1, 407 , 0.30


TBFIELD, XCOR, 145.38
TBFIELD, YCOR, -25.952
TBFIELD, ZCOR, 145
TBDATA, 1, 365 , 0.30


TBFIELD, XCOR, 140.79
TBFIELD, YCOR, -35.642
TBFIELD, ZCOR, 135
TBDATA, 1, 351 , 0.30


TBFIELD, XCOR, 140.79
TBFIELD, YCOR, -35.642
TBFIELD, ZCOR, 145
TBDATA, 1, 222 , 0.30


TBIN, ALGO, LMUL


TB, ELAS, 6
TBFIELD, XCOR, 191.75
```

```
TBFIELD, YCOR, -22.951
TBFIELD, ZCOR, 135
TBDATA, 1, 111 , 0.30


TBFIELD, XCOR, 185.36
TBFIELD, YCOR, -33.449
TBFIELD, ZCOR, 135
TBDATA, 1, 134 , 0.30


TBFIELD, XCOR, 185.36
TBFIELD, YCOR, -33.449
TBFIELD, ZCOR, 145
TBDATA, 1, 120 , 0.30


TBFIELD, XCOR, 177.85
TBFIELD, YCOR, -44.736
TBFIELD, ZCOR, 135
TBDATA, 1, 83 , 0.30


TBFIELD, XCOR, 177.85
TBFIELD, YCOR, -44.736
TBFIELD, ZCOR, 145
TBDATA, 1, 132 , 0.30


TBFIELD, XCOR, 177.05
TBFIELD, YCOR, -18.441
TBFIELD, ZCOR, 135
TBDATA, 1, 469 , 0.30


TBFIELD, XCOR, 177.05
TBFIELD, YCOR, -18.441
TBFIELD, ZCOR, 145
TBDATA, 1, 604 , 0.30


TBFIELD, XCOR, 171.15
TBFIELD, YCOR, -30.939
TBFIELD, ZCOR, 135
TBDATA, 1, 191 , 0.30
```

```
TBFIELD, XCOR, 171.15
TBFIELD, YCOR, -30.939
TBFIELD, ZCOR, 145
TBDATA, 1, 68 , 0.30


TBFIELD, XCOR, 166.77
TBFIELD, YCOR, -42.402
TBFIELD, ZCOR, 135
TBDATA, 1, 160 , 0.30


TBFIELD, XCOR, 166.77
TBFIELD, YCOR, -42.402
TBFIELD, ZCOR, 145
TBDATA, 1, 125 , 0.30


TBFIELD, XCOR, 160.38
TBFIELD, YCOR, -15.732
TBFIELD, ZCOR, 135
TBDATA, 1, 602 , 0.30


TBFIELD, XCOR, 160.38
TBFIELD, YCOR, -15.732
TBFIELD, ZCOR, 145
TBDATA, 1, 357 , 0.30


TBFIELD, XCOR, 155.90
TBFIELD, YCOR, -29.259
TBFIELD, ZCOR, 135
TBDATA, 1, 192 , 0.30


TBFIELD, XCOR, 155.90
TBFIELD, YCOR, -29.259
TBFIELD, ZCOR, 145
TBDATA, 1, 393 , 0.30


TBFIELD, XCOR, 151.25
TBFIELD, YCOR, -40.088
```

```
TBFIELD, ZCOR, 135
TBDATA, 1, 90 , 0.30

TBFIELD, XCOR, 151.25
TBFIELD, YCOR, -40.088
TBFIELD, ZCOR, 145
TBDATA, 1, 158 , 0.30

TBFIELD, XCOR, 149.06
TBFIELD, YCOR, -15.388
TBFIELD, ZCOR, 135
TBDATA, 1, 403 , 0.30

TBFIELD, XCOR, 149.06
TBFIELD, YCOR, -15.388
TBFIELD, ZCOR, 145
TBDATA, 1, 659 , 0.30

TBFIELD, XCOR, 145.38
TBFIELD, YCOR, -25.952
TBFIELD, ZCOR, 135
TBDATA, 1, 407 , 0.30

TBFIELD, XCOR, 145.38
TBFIELD, YCOR, -25.952
TBFIELD, ZCOR, 145
TBDATA, 1, 365 , 0.30

TBFIELD, XCOR, 140.79
TBFIELD, YCOR, -35.642
TBFIELD, ZCOR, 135
TBDATA, 1, 351 , 0.30

TBFIELD, XCOR, 140.79
TBFIELD, YCOR, -35.642
TBFIELD, ZCOR, 145
```

```
TBDATA, 1, 222 , 0.30

TBIN, ALGO, NNEI

ALLSEL
NSLE, S
*GET, NUM_N, NODE, 0, COUNT          ! GET NUMBER OF NODES
*GET, N_MIN, NODE, 0, NUM, MIN       ! GET MIN NODE NUMBER
*DO, I, 1, NUM_N, 1                   ! OUTPUT TO ASCII BY LOOPING
                                      OVER NODES

  CURR_N = N_MIN
  DZZ = NZ(CURR_N) - 150.45
  D, CURR_N, UZ, DZZ
  D, CURR_N, UX, 0
  D, CURR_N, UY, 0
  *GET, N_MIN, NODE, CURR_N, NXTH
*ENDDO
FINISH
/SOLU
ERESX, NO
NSUBST, 1, 1, 1
SOLV
FINISH
```

Appendix 2: Input File of the Anisotropic Femur Model in Section 4.2

```
/COM, download iso_femur.cdb from
/COM, www.feabea.net/models/iso_femur.cdb
/PREP7
CDREAD, DB, 'iso_femur', 'cdb'
ALLSEL
FINISH
/SOLU
TIME, 1
NSUBST, 1, 1, 1
OUTRES, ALL, ALL
SOLVE
FINISH

/POST1
*GET, NUMELEM, ELEM, 0, COUNT
*DIM, ARRAYS, ARRAY, NUMELEM, 6, 1, , ,
*DIM, DIRC, ARRAY, NUMELEM, 9, 1, , ,
*DIM, SPRINC, ARRAY, NUMELEM, 3, 1, , ,
*DIM, DIRE, ARRAY, NUMELEM, 6, 1, , ,

SET, FIRST

AVPRIN, 0, ,
ETABLE, S_1, S, 1
AVPRIN, 0, ,
ETABLE, S_2, S, 2
AVPRIN, 0, ,
ETABLE, S_3, S, 3
*VGET, SPRINC(1, 1), ELEM, 1, ETAB, S_1, , , 2      !GET PRINCIPAL
                                                    STRESSES
```

```
*VGET, SPRINC(1, 2), ELEM, 1, ETAB, S_2, , , 2
*VGET, SPRINC(1, 3), ELEM, 1, ETAB, S_3, , , 2
*VFUN, DIRC(1, 1), DIRCOS, ARRAYS(1, 1)

*DO, I, 1, NUMELEM, 1
  *IF, SPRINC(I, 1), ABGT, SPRINC(I, 3), THEN
        DIRE(I, 1) = DIRC(I, 1)
        DIRE(I, 2) = DIRC(I, 2)
        DIRE(I, 3) = DIRC(I, 3)
   *ELSE
        DIRE(I, 1) = DIRC(I, 7)
        DIRE(I, 2) = DIRC(I, 8)
        DIRE(I, 3) = DIRC(I, 9)
   *ENDIF
   DIRE(I, 4) = DIRC(I, 4)
   DIRE(I, 5) = DIRC(I, 5)
   DIRE(I, 6) = DIRC(I, 6)
*ENDDO
FINISH

/PREP7
*DO, NN, 1, 280
  *GET, EE, EX, NN

  DENS2 = (EE/7607)**(1/1.853)
  DENS3 = (EE/7607)**(1/1.853)
  ! CORTICAL BONE
  CTE2 = 2314*DENS2**1.57
  CTE3 = 2314*DENS2**1.57
  CTE1 = 2065*DENS2**3.09
  CT23 = 0.25
  CT12 = 0.32
  CT13 = 0.32
  GT23 = CTE1/2/(1 + CT12)
  GT12 = 3300
  GT13 = 3300
  ! CANCELLOUS BONE
```

```
CAE2 = 1157*DENS3**1.78
CAE3 = 1157*DENS3**1.78
CAE1 = 1094*DENS3**1.64
CA23 = 0.25
CA12 = 0.32
CA13 = 0.32
GA23 = CAE1/2/(1 + CA12)
GA12 = 110
GA13 = 110
! DEFINE MATERIAL ID FOR CORTICAL BONE
MP, EX, 1000 + NN, CTE1
MP, EY, 1000 + NN, CTE2
MP, EZ, 1000 + NN, CTE3
MP, GXY, 1000 + NN, GT12
MP, GYZ, 1000 + NN, GT23
MP, GXZ, 1000 + NN, GT13
MP, PRXY, 1000 + NN, CT12
MP, PRYZ, 1000 + NN, CT23
MP, PRXZ, 1000 + NN, CT13
! DEFINE MATERIAL ID FOR CANCELLOUS BONE
MP, EX, 2000 + NN, CAE1
MP, EY, 2000 + NN, CAE2
MP, EZ, 2000 + NN, CAE3
MP, GXY, 2000 + NN, GA12
MP, GYZ, 2000 + NN, GA23
MP, GXZ, 2000 + NN, GA13
MP, PRXY, 2000 + NN, CA12
MP, PRYZ, 2000 + NN, CA23
MP, PRXZ, 2000 + NN, CA13


*ENDDO
ALLSEL

*GET, NUM_E, ELEM, 0, COUNT ! GET NUMBER OF NODES
*GET, N_MIN, ELEM, 0, NUM, MIN ! GET MIN NODE NUMBER
*DO, I, 1, NUM_E, 1
  CURR_N = N_MIN
```

```
*GET, ET, TYPE, CURR_N

*GET, IMAT, MAT, CURR_N

*IF, ET, GT, 2, THEN

     EMODIF, CURR_N, MAT, 2000 + IMAT

*ELSE

     EMODIF, CURR_N, MAT, 1000 + IMAT

 *ENDIF


 *GET, N_MIN, ELEM, CURR_N, NXTH

*ENDDO


*DO, E, 1, NUMELEM, 1 !ASSIGN PRINCIPAL DIRECTIONS TO THE ELEMENT
COORDINATE SYSTEM

  ICOR = 2E5 + E

  CSYS, 0

  ESEL, S, ELEM, , E

  NSLE, S

  X_ = CENTRX(E)

  Y_ = CENTRY(E)

  Z_ = CENTRZ(E)

  CSYS, 4

  AA = ABS(DIRE(E, 1)) + ABS(DIRE(E, 2)) + ABS(DIRE(E, 3))

  *IF, AA, GT, 0.1, THEN

WPLANE, 1, X_, Y_, Z_, X_ + DIRE(E, 1), Y_ + DIRE(E, 2),
  Z_ + DIRE(E, 3), X_ + DIRE(E, 4), Y_ + DIRE(E, 5), Z_ + DIRE(E, 6)

   CSWPLA, ICOR, 0

   EMODIF, E, ESYS, ICOR

  *ENDIF

*ENDDO


ALLSEL

SAVE, 'ANISO_FEMUR', 'DB'

FINISH
```

Appendix 3: Input File of the XFEM Crack-Growth Model in Section 5.2

```
/COM, this model can be repeated in ANSYS190 and later versions
/PREP7
ET, 1, PLANE182
KEYOPT, 1, 3, 2

A = 0.25
B = 0.3
RC = 0.03
R = 0.1
RT = 0.1 + 0.003
K, 1, 0.35356, 0.01238 + A
K, 2, 0.61337, -0.13762 + A
K, 3, 0.67587, -0.029367 + A
K, 4, 0.41606, 0.12063 + A
A, 1, 2, 3, 4

RECTNG, 0, 3*B, 0, 2*A

CYL4, 0.5*B, 0.5*A, RC
CYL4, 0.5*B, 0.5*A, R
CYL4, 0.5*B, 0.5*A, RT
CYL4, 2.6*B, 1.5*A, RC
CYL4, 2.6*B, 1.5*A, R
CYL4, 2.6*B, 1.5*A, RT

AOVLAP, ALL
LESIZE, 3, , , 21, , , , , 1
LESIZE, 4, , , 15, , , , , 1
```

```
LESIZE, 6, , , 10, , , , , 1
LESIZE, 8, , , 10, , , , , 1
LESIZE, 5, , , 16, , , , , 1
LESIZE, 7, , , 16, , , , , 1
MSHAPE, 0, 2D
MSHKEY, 1
AMESH, 1

MSHKEY, 0
ASEL, U, AREA, , 1
AMESH, ALL

! ELEMENT COMPONENT REQUIRED FOR XFENRICH COMMAND
ESEL, ALL
CM, TESTCMP, ELEM
ALLSEL

! MATERIAL 1 INTERSTITIAL
MP, EX, 1, 14000
MP, NUXY, 1, 0.15

! MATERIAL 2 OSTEON
MP, EX, 2, 7000
MP, NUXY, 2, 0.17

! MATERIAL 3 CEMENT
MP, EX, 3, 5000
MP, NUXY, 3, 0.49

!CRACK GROWTH CRITERION
TB, CGCR, 2, , , STTMAX
TBDATA, 1, 10.5

! INTERFACE BEHAVIOR
TB, CGCR, 2, , , RLIN
TBDATA, 1, 0.04, , 0.2
```

```
ASEL, S, AREA, , 10, 12, 2
ESLA, S
EMODIF, ALL, MAT, 3
ALLSEL


ASEL, S, AREA, , 3, 9, 3
ASEL, A, AREA, , 11
ESLA, S
EMODIF, ALL, MAT, 2
ALLSEL


! MESH THE CRACK WITH MESH200 ELEMENTS
ET, 2, 200, 0                              ! 2D LINE WITH 2 NODES
K, 110, (0.48170 + 0.48587)/2, (0.25575 + 0.26297)/2
K, 111, (0.53119 + 0.53536)/2, (0.22718 + 0.23440)/2
L, 110, 111
TYPE, 2
LMESH, 33


ALLSEL


! ELEMENT COMPONENT FOR MESH200 ELEMENTS
ESEL, S, TYPE, , 2
CM, M200EL, ELEM
ALLSEL


! NODE COMPONENT FOR CRK FRONT
ESEL, S, TYPE, , 2
NSLE, S
NSEL, U, LOC, X, 0.49, 0.52
CM, M200ND, NODE
ALLSEL


!DEFINE ENRICHMENT IDENTIFICATION
XFENRICH, ENRICH1, TESTCMP       ! FOR PHANTOM-NODE METHOD
```

```
!DEFINE CRACK DATA
XFCRKMESH, ENRICH1, M200EL, M200ND
XFLIST
ALLSEL

NSEL, S, LOC, X, 0
D, ALL, UX
NSEL, S, LOC, Y, 0
D, ALL, UY
NSEL, S, LOC, Y, 2*A
SF, ALL, PRES, -10
ALLSEL
WPSTYLE, , , , , , , , 1
WPROTA, 150, ,
CSWPLA, 12, 0
WPROTA, -150

WPSTYLE, , , , , , , , 1
WPROTA, -30, ,
CSWPLA, 11, 0
CSYS, 0

FINISH

/SOLU
ANTYPE, 0
TIME, 1.0
DELTIM, 0.025, 1.0E-03, 0.025
OUTRES, ALL, ALL
NCNV, , , 100
/NERR, , , -1
!CINT CALCULATIONS
CINT, NEW, 1
CINT, CXFE, _XFCRKFREL1
CINT, TYPE, STTMAX
CINT, NCON, 6
```

```
CINT, NORM, 12, 2
CINT, RSWEEP, 181, -90, 90

!CGROW CALCULATIONS
CGROW, NEW, 1
CGROW, CID, 1
CGROW, METHOD, XFEM
CGROW, FCOPTION, MTAB, 2

!CINT CALCULATIONS
CINT, NEW, 2
CINT, CXFE, _XFCRKFREL2
CINT, TYPE, STTMAX
CINT, NCON, 6
CINT, NORM, 11 , 2
CINT, RSWEEP, 181, -90, 90

!CGROW CALCULATIONS
CGROW, NEW, 2
CGROW, CID, 2
CGROW, METHOD, XFEM
CGROW, FCOPTION, MTAB, 2
XFPR, 1

SOLVE
```

Appendix 4: Input File of the Abdominal Aortic Aneurysm Model in Section 7.2

```
/COM, download aaa_model.dat from
/COM, www.feabea.net/models/aaa_model.dat
/PREP7
/INP, aaa_model, dat
ALLSEL

*DIM, _LOADVARI36, TABLE, 2, 1, 1, TIME,
! TIME VALUES
_LOADVARI36(1, 0, 1) = 0.
_LOADVARI36(2, 0, 1) = 1.
! LOAD VALUES
_LOADVARI36(1, 1, 1) = 0.
_LOADVARI36(2, 1, 1) = 1.4665E-002

ESEL, ALL
! ODGEN MATERIAL DEFINITION
TB, HYPE, 1, 1, 1, OGDEN
TBDATA, 1, 0.00500258144460596, 46.8776155527868, 0

SECTYPE, 1, SHELL
SECDATA, 1.5
SECOFF, TOP

MAT, 4
R, 12
REAL, 12
ET, 11, 170
ET, 12, 175
```

```
KEYOPT, 12, 12, 5
KEYOPT, 12, 4, 1
KEYOPT, 12, 2, 2
KEYOPT, 11, 2, 1
KEYOPT, 11, 4, 111111
TYPE, 12
NSEL, S, LOC, Y, -50
ESLN, S, 0
ESURF

TYPE, 11
! CREATE A PILOT NODE
N, 10004, -7.902, -50, 0
TSHAP, PILO
E, 10004
! GENERATE THE CONTACT SURFACE

D, 10004, ALL
ALLSEL

MAT, 5
R, 13
REAL, 13
ET, 13, 170
ET, 14, 175
KEYOPT, 14, 12, 5
KEYOPT, 14, 4, 1
KEYOPT, 14, 2, 2
KEYOPT, 13, 2, 1
KEYOPT, 13, 4, 111111
TYPE, 14
NSEL, S, LOC, Y, 53.78
ESLN, S, 0
ESURF

TYPE, 13
! CREATE A PILOT NODE
```

```
N, 10005, -7.902, 53.78, 0
TSHAP, PILO
E, 10005
! GENERATE THE CONTACT SURFACE

D, 10005, ALL
ALLSEL

FINI

/COM, *******************************************************
/COM, **********************  SOLUTION  ********************
/COM, *******************************************************
/SOLU
ANTYPE, 0          ! STATIC ANALYSIS
NLGEOM, ON         ! TURN ON LARGE DEFORMATION EFFECTS
ESEL, S, TYPE, , 2
NSLE
SF, ALL, PRES, %_LOADVARI36%
ALLSEL

NSUB, 50, 1000, 25
TIME, 1.
OUTRES, ALL, ALL

SOLVE
```

Appendix 5: Input File of the Periodontal Ligament Creep Model in Section 8.2

```
/PREP7
/COM, ********** MATERIAL DEFINITION **********
MP, EX, 2, 0.23,
MP, NUXY, 2, 0.49,


TB, PRONY, 2, , 3, SHEAR        ! DEFINE VISCOUSITY PARAMETERS
TBDATA, 1, 0.91, 0.0025, 0.05, 0.1, 0.005, 0.5
TB, PRONY, 2, , 3, BULK         ! DEFINE VISCOUSITY PARAMETERS
TBDATA, 1, 0.155, 0.0025, 0.4, 0.1, 0.15, 0.5


MP, EX, 4, 13700               ! CORTICAL
MP, NUXY, 4, 0.3,


MP, EX, 3, 1370
MP, NUXY, 3, 0.3,             ! CANCELLOUS


MP, EX, 1, 13700             ! TOOTH
MP, NUXY, 1, 0.3,


K, 1, -3
K, 2, -2, -9
K, 3, 0, -18
K, 4, 2, -9
K, 5, 3
BSPLIN, 1, 2, 3, 4, 5


K, 6, -3.2
K, 7, -2.2, -9
```

```
K, 8, 0, -18.2
K, 9, 2.2, -9
K, 10, 3.2
BSPLIN, 6, 7, 8, 9, 10
K, 11,
K, 12, 0, 0, 1
L, 1, 6
L, 5, 10
L, 1, 5
AL, 1, 2, 3, 4
AL, 1, 5
K, 13, 0, -25
L, 11, 13
ASBL,    2,    6
K, 15, 0, -26
L, 11, 15
ASBL,    1,    1
ADELE, 4, 5
K, 16, 0, 1
VROTAT, 2, 3, , , , , 11, 16
WPROTAT, , -90
CYL4, 0, 0, 3, , , , 4
WPROTA, , 90
BLOCK, -10, 10, -2, -30, -5, 5,
BLOCK, -10, 10, 0, -2, -5, 5,
VOVLAP, ALL,
ET, 1, 187
MAT, 2
ESIZE, 0.5
VMESH, 20, 27
MAT, 4
ESIZE, 3
VMESH, 30
MAT, 3
ESIZE, 3
VMESH, 29
```

```
ALLSEL
VSEL, U, VOLU, , 29, 30
VSEL, U, VOLU, , 20, 27
MAT, 1
ESIZE, 0.5
VMESH, ALL

ALLSEL

NSEL, S, LOC, Y, -30
D, ALL, ALL
ALLSEL

NSEL, S, LOC, Z, 5
NSEL, A, LOC, Z, -5
D, ALL, UZ, 0
NSEL, S, LOC, X, 10
NSEL, A, LOC, X, -10
D, ALL, UX, 0
ALLSEL

NSEL, S, LOC, Y, 4
F, ALL, FZ, 0.0025
ALLSEL
FINISH
/SOLU
TIME, 2.5
NLGEOM, ON
KBC, 1
NSUBST, 20, 1000, 10
OUTRES, ALL, ALL
SOLV
TIME, 6
NSEL, S, LOC, Y, 4
F, ALL, FZ, 0
ALLSEL
```

```
SOLV
FINISH

/POST26
NSOL, 2, 47680, U, Z,
PLVAR, 2, , , , , , , , , ,
```

Appendix 6: Input File of the Intervertebral Disc Model with Fiber Enhancement in Section 9.1.2

```
/PREP7
ANGLE = 30
ANGLE1 = 180 - ANGLE
! CROSS AREA
AA = 1E-4
! DISTANCES BETWEEN FIBERS
SS = 0.01

K, 1, 0, 0
K, 2, -20, 16
K, 3, -11, 26
K, 4, 0, 23
BSPLIN, 1, 2, 3, 4, , , 1, 0, 0, 0.966, -0.259, 0.0
K, 11, 0, 5
K, 12, -11, 14
K, 13, -8, 20
K, 14, 0, 18
BSPLIN, 11, 12, 13, 14, , , 1, 0, 0, 0.966, -0.259, 0.0

L, 4, 14
L, 11, 14
L, 1, 11
ALLSEL

AL, 1, 2, 3, 5
AL, 2, 4
```

```
ET, 1, 182
ET, 2, 185
ET, 3, 80

MP, EX, 1, 1.0           ! ANNULUS
MP, NUXY, 1, 0.48

MP, EX, 2, 0.4
MP, ALPX, 2, 0.001

MP, EX, 3, 223.8         ! ENDPLATE
MP, NUXY, 3, 0.4

MP, EX, 4, 156           ! MAT4 – MAT18 ARE FIBER MATERIAL PROPERTIES
MP, NUXY, 4, 0.1

MP, EX, 5, 148
MP, NUXY, 5, 0.1

MP, EX, 6, 140
MP, NUXY, 6, 0.1

MP, EX, 7, 132
MP, NUXY, 7, 0.1

MP, EX, 8, 124
MP, NUXY, 8, 0.1

MP, EX, 9, 116
MP, NUXY, 9, 0.1

MP, EX, 10, 108
MP, NUXY, 10, 0.1

MP, EX, 11, 100
MP, NUXY, 11, 0.1
```

```
MP, EX, 12, 92
MP, NUXY, 12, 0.1

MP, EX, 13, 84
MP, NUXY, 13, 0.1

MP, EX, 14, 76
MP, NUXY, 14, 0.1

MP, EX, 15, 68
MP, NUXY, 15, 0.1

MP, EX, 16, 60
MP, NUXY, 16, 0.1

MP, EX, 17, 52
MP, NUXY, 17, 0.1

MP, EX, 18, 44
MP, NUXY, 18, 0.1

! FIBER SECTION DATA
! AA = 1E-4; SS = 0.01
SECT, 1, REINF, SMEAR
SECD, 4, AA, SS, , ANGLE, ELEF, 2, 0
! TENSION ONLY
SECC, 1

SECT, 2, REINF, SMEAR
SECD, 5, AA, SS, , ANGLE1, ELEF, 2, 0
SECC, 1

SECT, 3, REINF, SMEAR
SECD, 6, AA, SS, , ANGLE, ELEF, 2, 0
SECC, 1
```

```
SECT, 4, REINF, SMEAR
SECD, 7, AA, SS, , ANGLE1, ELEF, 2, 0
SECC, 1


SECT, 5, REINF, SMEAR
SECD, 8, AA, SS, , ANGLE, ELEF, 2, 0
SECC, 1


SECT, 6, REINF, SMEAR
SECD, 9, AA, SS, , ANGLE1, ELEF, 2, 0
SECC, 1


SECT, 7, REINF, SMEAR
SECD, 10, AA, SS, , ANGLE, ELEF, 2, 0
SECC, 1


SECT, 8, REINF, SMEAR
SECD, 11, AA, SS, , ANGLE1, ELEF, 2, 0
SECC, 1


SECT, 9, REINF, SMEAR
SECD, 12, AA, SS, , ANGLE, ELEF, 2, 0
SECC, 1


SECT, 10, REINF, SMEAR
SECD, 13, AA, SS, , ANGLE1, ELEF, 2, 0
SECC, 1


SECT, 11, REINF, SMEAR
SECD, 14, AA, SS, , ANGLE, ELEF, 2, 0
SECC, 1


SECT, 12, REINF, SMEAR
SECD, 15, AA, SS, , ANGLE1, ELEF, 2, 0
SECC, 1
```

```
SECT, 13, REINF, SMEAR
SECD, 16, AA, SS, , ANGLE, ELEF, 2, 0
SECC, 1


SECT, 14, REINF, SMEAR
SECD, 17, AA, SS, , ANGLE1, ELEF, 2, 0
SECC, 1


SECT, 15, REINF, SMEAR
SECD, 18, AA, SS, , ANGLE, ELEF, 2, 0
SECC, 1


LESIZE, 1, , , 44, , , , , 1
LESIZE, 3, , , 15, , , , , 1
LESIZE, 4, , , 30
MAT, 1
MSHKEY, 1
AMESH, 1
MAT, 2
MSHKEY, 0
AMESH, 2


TYPE, 2
ESIZE, , 40
MAT, 1
VEXT, ALL, , , 0, 0, 11
ALLSEL, ALL


ESEL, S, ENAME, , 182
ACLEAR, 1, 2, , ,
ALLSEL, ALL


VSEL, S, VOLU, , 1              ! ANNULUS
ESLV, S
EMODIF, ALL, MAT, 1
ALLSEL
```

```
VSEL, S, VOLU, , 2              ! NUCLEUS
ESLV, S
EMODIF, ALL, MAT, 2
EMODIF, ALL, TYPE, 3
ALLSEL


NSEL, S, LOC, Z, 9.5, 11
ESLN, S
EMODIF, ALL, TYPE, , 2
EMODIF, ALL, MAT, 3            ! END PLATE
ALLSEL


NSEL, S, LOC, Z, 0, 1.5
ESLN, S
EMODIF, ALL, TYPE, , 2
EMODIF, ALL, MAT, 3            ! END PLATE
ALLSEL


! ELEMENTS FOR FIBER
*DO, J, 1, 15

  SECN, J
  ESEL, S, ELEM, , 1079 + (J - 1)*40, 1106 + (J - 1)*40

  *DO, I, 1, 43
    ESEL, A, ELEM, , 1079 + (J - 1)*40 + I*600, 1106 + I*600 + (J - 1)*40
  *ENDDO

  EREINF, ALL
  ALLSEL

*ENDDO
ALLSEL


NSEL, S, LOC, Z, 0
D, ALL, ALL
```

```
ALLSEL
FINISH

/SOLVE
NROPT, FULL
NSEL, S, LOC, X, 0
D, ALL, UX, 0
ALLSEL
BFE, ALL, TEMP, 1, 100     ! UNIFORM BODY FORCE ACROSS ALL ELEMENTS
TIME, 1
NSUB, 50, 1000, 20
OUTRES, ALL, ALL

SOLVE

/POST1
ESEL, TYPE, 4
ETAB, N11, S, X

/OUT
PLETAB, N11
PRETAB, N11
/DEVICE, VECTOR, 0
PLETAB, N11, NOAV
FINISH
```

Appendix 7: Input File of the Intervertebral Disc Model with Mesh Independent Fiber Enhancement in Section 9.2.2

```
/COM, download fiber_indep.iges from
/COM, www.feabea.net/models/fiber_indep.iges
/COM, this model can be repeated in ANSYS190 and later versions
/PREP7
CDOPT, IGES
CDREAD, SOLID, ", ", , 'fiber_indep', 'iges'


ANGLE = 30
ANGLE1 = 180 - ANGLE
! CROSS AREA
AA = 1E-4
! DISTANCES BETWEEN FIBERS
SS = 0.01


ET, 1, 182
ET, 4, 285


MP, EX, 1, 1.0          ! ANNULUS
MP, NUXY, 1, 0.48


MP, EX, 2, 0.4
MP, ALPX, 2, 0.001


MP, EX, 3, 223.8        ! ENDPLATE
MP, NUXY, 3, 0.4
```

```
MP, EX, 4, 156        ! MAT4-MAT18 ARE FIBER MATERIAL PROPERTIES
MP, NUXY, 4, 0.1

MP, EX, 5, 148
MP, NUXY, 5, 0.1

MP, EX, 6, 140
MP, NUXY, 6, 0.1

MP, EX, 7, 132
MP, NUXY, 7, 0.1

MP, EX, 8, 124
MP, NUXY, 8, 0.1

MP, EX, 9, 116
MP, NUXY, 9, 0.1

MP, EX, 10, 108
MP, NUXY, 10, 0.1

MP, EX, 11, 100
MP, NUXY, 11, 0.1

MP, EX, 12, 92
MP, NUXY, 12, 0.1

MP, EX, 13, 84
MP, NUXY, 13, 0.1

MP, EX, 14, 76
MP, NUXY, 14, 0.1

MP, EX, 15, 68
MP, NUXY, 15, 0.1

MP, EX, 16, 60
MP, NUXY, 16, 0.1
```

```
MP, EX, 17, 52
MP, NUXY, 17, 0.1

MP, EX, 18, 44
MP, NUXY, 18, 0.1

TYPE, 4
ESIZE, 2
VMESH, ALL

ET, 5, 200, 6
TYPE, 5
! AA = 1E-4; SS = 0.01

*DO, J, 1, 15, 2
  SECT, J, REINF, SMEAR
  SECD, J + 3, AA, SS, , ANGLE, MESH
  ! TENSION ONLY
  SECC, 1
  ASEL, S, AREA, , 9 + J
  ESIZE, 1
  SECN, J
  AMESH, 9 + J
  ALLSEL
*ENDDO

*DO, J, 2, 14, 2
  SECT, J, REINF, SMEAR
  SECD, J + 3, AA, SS, , ANGLE1, MESH
  ! TENSION ONLY
  SECC, 1
  ASEL, S, AREA, , 9 + J
  ESIZE, 1
  SECN, J
  AMESH, 9 + J
  ALLSEL
```

```
*ENDDO
ESEL, ALL
CSYS, 0
EREINF

VSEL, S, VOLU, , 3, 4
ESLV, S
EMODIF, ALL, MAT, 3      ! END PLATE
ALLSEL

VSEL, S, VOLU, , 7, 8
ESLV, S
EMODIF, ALL, MAT, 3      ! END PLATE
ALLSEL

VSEL, S, VOLU, , 5       ! ANNULUS
ESLV, S
EMODIF, ALL, MAT, 1
ALLSEL

VSEL, S, VOLU, , 6       ! NUCLEUS
ESLV, S
EMODIF, ALL, MAT, 2
ALLSEL

NSEL, S, LOC, Z, 1.23
D, ALL, ALL
ALLSEL
FINISH

/SOLVE
NROPT, FULL
NSEL, S, LOC, X, 0
D, ALL, UX, 0
ALLSEL
BFE, ALL, TEMP, 1, 100      ! UNIFORM BODY FORCE ACROSS ALL ELEMENTS
```

```
TIME, 1
NSUB, 50, 1000, 20
OUTRES, ALL, LAST

SOLVE

/POST1
ESEL, TYPE, 6
ETAB, N11, S, X

/OUT
PLETAB, N11
PRETAB, N11
/DEVICE, VECTOR, 0
!PLETAB, N11, NOAV
PLESOL, S, X
FINISH
```

Appendix 8: Input File of the Anterior Cruciate Ligament Model in Section 9.3.2

```
/COM,  download acl_model.db from
/COM,  www.feabea.net/models/acl_model.db
/COM,  acl_model.db can be resumed in ANSYS190 and later versions
/PREP7
RESUME, 'acl_model', 'db'
/COM,  * ANISOTROPIC HYPERELASTIC MATERIAL PROPERTIES
A1 = 1.5          ! C1 FROM PENA PAPER
A2 = 0            ! C2 FROM PENA PAPER
A3 = 0
B1 = 0
B2 = 0
B3 = 0
C1 = 4.39056     ! C3 FROM PENA PAPER
C2 = 12.1093     ! C4 FROM PENA PAPER
E1 = 0
E2 = 0


! THE FOLLOWING MATERIAL PROPERTIES DEFINED IN DB FILE
!TB, AHYPER, 1, , 10, EXP
!TBDATA, 1,  A1, A2, A3, B1, B2, B3
!TBDATA, 7,  C1, C2, E1, E2


!TB, AHYPER, 1, , , AVEC    ! ALIGNED WITH UNIAXIAL STRAIN DIRECTION
!TBDATA, 1,  1,  0,  0


!TB, AHYPER, 1, , , PVOL
!TBDATA, 1,  1E-3
```

```
!MP, EX, 2, 1000
!MP, NUXY, 2, 0.3

ESEL, S, TYPE, , 1
EMODIF, ALL, MAT, 1
ESEL, S, TYPE, , 2
EMODIF, ALL, MAT, 2
ALLSEL

ET, 12, 170
ET, 13, 175
KEYOPT, 13, 10, 2
KEYOPT, 13, 4, 2
KEYOPT, 13, 2, 2
KEYOPT, 13, 12, 5
R, 13

!PILOT NODE
TSHAP, PILOT
N, 100000, 72, 124, 50
TYPE, 12
REAL, 13
E, 100000

TYPE, 13
REAL, 13

CMSEL, S, BOTTOM, ELEM
NSLE, S
ESURF
ALLSEL, ALL

D, 100000, ALL, 0
ALLSEL
```

```
ET, 22, 170
ET, 23, 175
KEYOPT, 23, 10, 2
KEYOPT, 23, 4, 2
KEYOPT, 23, 2, 2
KEYOPT, 23, 12, 5
R, 23

!PILOT NODE
TSHAP, PILOT
N, 100001, 95, 44, 36
TYPE, 22
REAL, 23
E, 100001

TYPE, 23
REAL, 23

ESEL, S, TYPE, , 2
CMSEL, U, BOTTOM, ELEM
NSLE, S
ESURF
ALLSEL, ALL

D, 100001, UX, 4
D, 100001, UY, 0
D, 100001, UZ, 0
ALLSEL

CS, 12, 0, 95123, 95128, 95078, 1, 1,
ESEL, S, TYPE, , 1
EMODIF, ALL, ESYS, 12
ALLSEL
CSYS, 0
FINISH
```

```
/SOLU

TIME, 1
NLGEOM, ON
NSUBST, 20, 1000, 10
SOLV
FINISH
```

Appendix 9: Input File of Subroutine UserHyper in Section 10.2

```
subroutine userhyper(
    &              prophy, incomp, nprophy, invar,
    &              potential, pinvder)
c*************************************************************
c
c   *** example of user hyperelastic routine
c
c       this example uses arruda hyperelasticity model
c       which is the same ansys tb,boyce
c
c   input arguments
c   ===============
c   prophy    (dp,ar(*),i)   material property array
c   nprophy   (int,sc,i)     # of material constants
c   invar     dp,ar(3)       invariants
c
c   output arguments
c   ================
c   incomp    (log,sc,i)     fully incompressible or compressible
c   potential dp,sc          value of potential
c   pinvder   dp,ar(10)      der of potential wrt i1,i2,j
c                        1 - der of potential wrt i1
c                        2 - der of potential wrt i2
c                        3 - der of potential wrt i1i1
c                        4 - der of potential wrt i1i2
c                        5 - der of potential wrt i2i2
c                        6 - der of potential wrt i1j
c                        7 - der of potential wrt i2j
c                        8 - der of potential wrt j
```

```fortran
c  _____                    9 - der of potential wrt jj
c
c***********************************************************
c
c --- parameters
c
#include "impcom.inc"
      double precision zero, one, two, three, half, toler
      parameter     (zero = 0.d00,
     &          one  = 1.0d0,
     &          half = 0.5d0,
     &          two  = 2.d0,
     &          three = 3.d0,
     &          toler = 1.0d-12)


c
c --- argument list
c
      integer       nprophy
      double precision prophy(*), invar(*),
     &          potential, pinvder(*)
      logical       incomp
c
c --- local variables
c
      double precision i1, jj, a1, g1, od1, j1

c
      i1  = invar(1)
      jj  = invar(3)
      a1  = prophy(1)
      g1  = prophy(2)
      od1 = prophy(3)
      potential = zero
      pinvder(1) = zero
      pinvder(3) = zero
```

```
potential = a1/two/g1 * (exp(g1 * (i1 - three)) - one)
pinvder(1) = a1/two*exp(g1 * (i1 - three))
pinvder(3) = g1*pinvder(1)

j1 = one / jj
pinvder(8) = zero
pinvder(9) = zero
if(od1 .gt. toler) then
  od1 = one / od1
  incomp = .false.
  potential = potential + od1 * ((jj*jj - one) * half - log(jj))
  pinvder(8) = od1 * (jj - j1)
  pinvder(9) = od1 * (one + j1 * j1)
end if

c


  return
  end
```

Appendix 10: Input File of the Head Impact Model in Section 11.2

```
/COM, download head.cdb from
/COM, www.feabea.net/models/head.cdb
/PREP7
CDREAD, DB, 'head', 'cdb'
ET, 1, 212
KEYOPT, 1, 3, 2
KEYOPT, 1, 12, 1
ET, 2, 182
KEYOPT, 2, 3, 2

MP, EX, 1, 6.67E4
MP, NUXY, 1, 0.48
MP, DENS, 1, 1040
FPX = 4.8E-8                    ! PERMEABILITY OF BRAIN TISSUES
TB, PM, 1, , , PERM
TBDATA, 1, FPX, FPX, FPX
TB, PM, 1, , , BIOT
TBDATA, 1, 1.0

MP, EX, 2, 6500E6
MP, NUXY, 2, 0.22
MP, DENS, 2, 1412

MP, EX, 3, 6.67E4
MP, NUXY, 3, 0.499
MP, DENS, 3, 1040
FPX = 1                        ! PERMEABILITY OF CSF
ONE = 1.0
TB, PM, 3, , , PERM
TBDATA, 1, FPX, FPX, FPX
```

```
TB, PM, 3, , , BIOT
TBDATA, 1, ONE

ESEL, S, MAT, , 2
EMODIF, ALL, TYPE, , 2
ALLSEL

NSEL, S, NODE, , 1, 3
NSEL, A, NODE, , 51, 53
NSEL, A, NODE, , 99, 100
D, ALL, UX, 0
D, ALL, UY, 0
ALLSEL

F, 82, FX, 8000              ! UPLOAD
ALLSEL

FINISH
/SOLU
ANTYPE, SOIL
KBC, 0
OUTRES, ALL, ALL
NLGEOM, ON
TIME, 1E-3
NSUBST, 100, 1000, 20
SOLV
TIME, 2E-3
F, 82, FX, 0
ALLSEL
SOLV
TIME, 100E-3               ! NO LOAD
SOLV
FINISH

/POST26
NSOL, 2, 582, PRES, ,
PLVAR, 2, , , , , , , , , , ,
```

Appendix 11: Input File of the Intervertebral Disc Model in Section 11.3

```
/PREP7

K, 1, 0, 0
K, 2, -20, 16
K, 3, -11, 26
K, 4, 0, 23
BSPLIN, 1, 2, 3, 4, , , 1, 0, 0, 0.966, -0.259, 0.0
K, 11, 0, 5
K, 12, -11, 14
K, 13, -8, 20
K, 14, 0, 18
BSPLIN, 11, 12, 13, 14, , , 1, 0, 0, 0.966, -0.259, 0.0

L, 4, 14
L, 11, 14
L, 1, 11
ALLSEL

AL, 1, 2, 3, 5
AL, 2, 4

ET, 1, 182
ET, 2, 215
KEYOPT, 2, 12, 1
ET, 3, 215
KEYOPT, 3, 12, 1
ET, 4, 185
```

```
E1 = 1.5
MP, EX, 1, E1                    ! YOUNG'S MODULUS, MPA
MP, NUXY, 1, 0.17               ! POISSON RATIO
FPX1 = 2E-4                      ! FLUID PERMEABILITY
TB, PM, 1, , , PERM
TBDATA, 1, FPX1, FPX1, FPX1

FPX2 = 1E-3                      ! NUCLEUS PERMEABILITY

MP, EX, 2, 0.75,                ! MPA
MP, NUXY, 2, 0.17,

TB, PM, 2, , , PERM
TBDATA, 1, FPX2, FPX2, FPX2
W = 1
CVC = E1*FPX1                    ! COEFFICIENT OF CONSOLIDATION
TV = 80                         ! TIME FACTOR STEP
TT = TV*2*W*2*W/CVC             ! CRITICAL TIME

MP, EX, 3, 223.8                ! ENDPLATE
MP, NUXY, 3, 0.4

LESIZE, 1, , , 44, , , , , , 1
LESIZE, 3, , , 8, , , , , 1
LESIZE, 4, , , 15
MAT, 1
MSHKEY, 1
AMESH, 1
MAT, 2
MSHKEY, 0
AMESH, 2

TYPE, 2

ESIZE, , 20
MAT, 1
VEXT, ALL, , , 0, 0, 11
ALLSEL, ALL
```

```
ESEL, S, ENAME, , 182
ACLEAR, 1, 2, , ,
ALLSEL, ALL

VSEL, S, VOLU, , 1            ! ANNULUS
ESLV, S
EMODIF, ALL, MAT, 1
ALLSEL

VSEL, S, VOLU, , 2            ! NUCLEUS
ESLV, S
EMODIF, ALL, MAT, 2
EMODIF, ALL, TYPE, 3
ALLSEL

NSEL, S, LOC, Z, 9.5, 11
ESLN, S
EMODIF, ALL, TYPE, 4
EMODIF, ALL, MAT, 3          ! END PLATE
ALLSEL

NSEL, S, LOC, Z, 0, 1.5
ESLN, S
EMODIF, ALL, TYPE, 4
EMODIF, ALL, MAT, 3          ! END PLATE
ALLSEL

ASEL, S, AREA, , 4
NSLA, S, 1
NSEL, R, LOC, Z, 1.5, 9.5
D, ALL, PRES, 0
ALLSEL

NSEL, S, LOC, Z, 0
D, ALL, ALL
ALLSEL
```

```
NSEL, S, LOC, Z, 11
F, ALL, FZ, -0.32
ALLSEL
FINISH

/SOLVE
NSEL, S, LOC, X, 0
D, ALL, UX, 0
ALLSEL
TIME, TT
NROPT, UNSYM
KBC, 1
NSUB, 50, 1000, 20
OUTRES, ALL, ALL

SOLVE
```

Appendix 12: Input File of the Knee Contact Model in Section 13.2

```
/COM, download knee_model.cdb from
/COM, www.feabea.net/models/knee_model.cdb
CDREAD, DB, 'knee_model', 'cdb',
/PREP7


! CONTACT BETWEEN TIBIA CARTILAGE AND FEMORAL CARTILAGE
MAT, 2
R, 2
REAL, 2
ET, 5, 170
ET, 6, 174


KEYOPT, 6, 9, 1
KEYOPT, 6, 10, 0
KEYOPT, 6, 12, 0
KEYOPT, 6, 5, 2
R, 2,
! GENERATE THE TARGET SURFACE
CMSEL, S, PFCS, NODE
TYPE, 5
ESLN, S, 0
ESURF
CMSEL, S, _ELEMCM
! GENERATE THE CONTACT SURFACE
CMSEL, S, PTCS, NODE
TYPE, 6
ESLN, S, 0
ESURF
ALLSEL
```

```
! CONTACT BETWEEN TIBIA AND MENISCUS
MAT, 2
R, 3
REAL, 3
ET, 7, 170
ET, 8, 174

KEYOPT, 8, 9, 1
KEYOPT, 8, 10, 0
KEYOPT, 8, 12, 0
KEYOPT, 8, 5, 2
R, 3,
! GENERATE THE TARGET SURFACE
CMSEL, S, TCS, NODE
TYPE, 7
ESLN, S, 0
ESURF
CMSEL, S, _ELEMCM
! GENERATE THE CONTACT SURFACE
CMSEL, S, MMTIBR, NODE
CMSEL, A, LMTIBR, NODE
TYPE, 8
ESLN, S, 0
ESURF

! CONTACT BETWEEN FEMUR AND MENISCUS
MAT, 2
R, 4
REAL, 4
ET, 9, 170
ET, 10, 174
KEYOPT, 10, 9, 1
KEYOPT, 10, 10, 0
KEYOPT, 10, 5, 2
R, 4,
! GENERATE THE TARGET SURFACE
CMSEL, S, FCS, NODE
```

```
TYPE, 9
ESLN, S, 0
ESURF
CMSEL, S, _ELEMCM
! GENERATE THE CONTACT SURFACE
CMSEL, S, MMFEMR, NODE
CMSEL, A, LMFEMR, NODE
TYPE, 10
ESLN, S, 0
ESURF

! CONTACT BETWEEN TIBIA AND CARTILAGE
MAT, 2
R, 5
REAL, 5
ET, 11, 170
ET, 12, 174
KEYOPT, 12, 9, 1
KEYOPT, 12, 12, 5
KEYOPT, 12, 5, 3
R, 5,
RMORE,
! GENERATE THE TARGET SURFACE
CMSEL, S, INTCS, NODE
TYPE, 11
ESLN, S, 0
ESURF
CMSEL, S, _ELEMCM
! GENERATE THE CONTACT SURFACE
CMSEL, S, TC2TIB, NODE
TYPE, 12
ESLN, S, 0
ESURF
ALLSEL

! CONTACT BETWEEN FEMUR AND CARTILAGE
MAT, 2
```

```
R, 6
REAL, 6
ET, 15, 170
ET, 16, 174
KEYOPT, 16, 9, 1
KEYOPT, 16, 12, 5
KEYOPT, 16, 5, 3
R, 6,
! GENERATE THE TARGET SURFACE
CMSEL, S, INF2FEM, NODE
TYPE, 15
ESLN, S, 0
ESURF
CMSEL, S, _ELEMCM
! GENERATE THE CONTACT SURFACE
CMSEL, S, F2FEM, NODE
TYPE, 16
ESLN, S, 0
ESURF
ALLSEL

A = 3E6
ESEL, S, TYPE, , 11

! SELECT PART

ET, 3, 185

*GET, NUM_E, ELEM, 0, COUNT      ! GET NUMBER OF ELEMENTS
*GET, E_MIN, ELEM, 0, NUM, MIN   ! GET MIN ELEMENT NUMBER

SHPP, OFF

*DIM, E1, ARRAY, NUM_E, 4

*DO, I, 1, NUM_E, 1        ! OUTPUT TO ASCII BY LOOPING OVER ELEMENTS
  CURR_E = E_MIN
  P1X = NX(NELEM(CURR_E, 1))
```

```
P1Y = NY(NELEM(CURR_E, 1))
P1Z = NZ(NELEM(CURR_E, 1))
P2X = NX(NELEM(CURR_E, 2))
P2Y = NY(NELEM(CURR_E, 2))
P2Z = NZ(NELEM(CURR_E, 2))
P3X = NX(NELEM(CURR_E, 3))
P3Y = NY(NELEM(CURR_E, 3))
P3Z = NZ(NELEM(CURR_E, 3))

PX = (P2Y - P1Y)*(P3Z - P2Z) - (P2Z - P1Z)*(P3Y - P2Y)
PY = (P2Z - P1Z)*(P3X - P2X) - (P2X - P1X)*(P3Z - P2Z)
PZ = (P2X - P1X)*(P3Y - P2Y) - (P2Y - P1Y)*(P3X - P2X)
PP = SQRT(PX*PX + PY*PY + PZ*PZ)
PXX = PX/PP
PYY = PY/PP
PZZ = PZ/PP
*DO, J, 1, 4
    XX = NX(NELEM(CURR_E, J)) - PXX
    YY = NY(NELEM(CURR_E, J)) - PYY
    ZZ = NZ(NELEM(CURR_E, J)) - PZZ
    N, A + (I - 1)*4 + J, XX, YY, ZZ
*ENDDO

TYPE, 3
MAT, 4
E, NELEM(CURR_E, 1), NELEM(CURR_E, 2), NELEM(CURR_E, 3), NELEM
(CURR_E, 4), A + (I - 1)*4 + 1, A + (I - 1)*4 + 2, A+(I - 1)*4 + 3, A +
(I - 1)*4 + 4

*GET, E_MIN, ELEM, CURR_E, NXTH
*ENDDO
ALLSEL

A = 4E6

ESEL, S, TYPE, , 15
```

```
! SELECT PART

ET, 3, 185

*GET, NUM_E, ELEM, 0, COUNT ! GET NUMBER OF ELEMENTS
*GET, E_MIN, ELEM, 0, NUM, MIN ! GET MIN ELEMENT NUMBER

SHPP, OFF

*DIM, E1, ARRAY, NUM_E, 4
*DO, I, 1, NUM_E, 1 ! OUTPUT TO ASCII BY LOOPING OVER ELEMENTS
  CURR_E = E_MIN
  P1X = NX(NELEM(CURR_E, 1))
  P1Y = NY(NELEM(CURR_E, 1))
  P1Z = NZ(NELEM(CURR_E, 1))
  P2X = NX(NELEM(CURR_E, 2))
  P2Y = NY(NELEM(CURR_E, 2))
  P2Z = NZ(NELEM(CURR_E, 2))
  P3X = NX(NELEM(CURR_E, 3))
  P3Y = NY(NELEM(CURR_E, 3))
  P3Z = NZ(NELEM(CURR_E, 3))

  PX = (P2Y - P1Y)*(P3Z - P2Z) - (P2Z - P1Z)*(P3Y - P2Y)
  PY = (P2Z - P1Z)*(P3X - P2X) - (P2X - P1X)*(P3Z - P2Z)
  PZ = (P2X - P1X)*(P3Y - P2Y) - (P2Y - P1Y)*(P3X - P2X)
  PP = SQRT(PX*PX + PY*PY + PZ*PZ)
  PXX = PX/PP
  PYY = PY/PP
  PZZ = PZ/PP
  *DO, J, 1, 4
      XX = NX(NELEM(CURR_E, J)) - PXX
      YY = NY(NELEM(CURR_E, J)) - PYY
      ZZ = NZ(NELEM(CURR_E, J)) - PZZ
      N, A + (I - 1)*4 + J, XX, YY, ZZ
  *ENDDO

  TYPE, 3
  MAT, 5
```

```
E, NELEM(CURR_E, 1), NELEM(CURR_E, 2), NELEM(CURR_E, 3), NELEM
(CURR_E, 4), A + (I - 1)*4 + 1, A + (I - 1)*4 + 2, A + (I - 1)*4 + 3, A +
(I - 1)*4 + 4

 *GET, E_MIN, ELEM, CURR_E, NXTH
*ENDDO
ALLSEL

MAT, 4
R, 15
REAL, 15
ET, 17, 170
ET, 18, 175
KEYOPT, 18, 12, 5
KEYOPT, 18, 4, 2
KEYOPT, 18, 2, 2
KEYOPT, 17, 2, 1
KEYOPT, 17, 4, 111111
TYPE, 17
! CREATE A PILOT NODE
N, 10000001, 80, 55, 55
TSHAP, PILO
E, 10000001
TYPE, 18
! GENERATE THE CONTACT SURFACE
NSEL, S, NODE, , 4E6, 5E6
ESLN, S, 0
ESURF

MAT, 4
R, 16
REAL, 16
ET, 19, 170
ET, 20, 175
KEYOPT, 20, 12, 5
KEYOPT, 20, 4, 2
KEYOPT, 20, 2, 2
KEYOPT, 19, 2, 1
```

```
KEYOPT, 19, 4, 111111
TYPE, 19
! CREATE A PILOT NODE
N, 10000002, 100, 95, 45
TSHAP, PILO
E, 10000002
TYPE, 20
! GENERATE THE CONTACT SURFACE
NSEL, S, NODE, , 3E6, 4E6
ESLN, S, 0
ESURF
ALLSEL
D, 10000001, UX, 0
D, 10000001, UY, 1
D, 10000001, UZ, 0
D, 10000001, ROTX, 0
D, 10000001, ROTY, 0
D, 10000001, ROTZ, 0
D, 10000002, ALL, 0
ALLSEL
CMSEL, S, MMANT, NODE
CMSEL, A, LMANT, NODE
CMSEL, A, MMPOST, NODE
CMSEL, A, LMPOST, NODE
D, ALL, ALL
ALLSEL

MP, EX, 4, 20E3
MP, NUXY, 4, 0.3

MP, EX, 5, 20E3
MP, NUXY, 5, 0.3

MP, EX, 3, 5
MP, NUXY, 3, 0.46

MP, EX, 6, 5
```

```
MP, NUXY, 6, 0.46

MP, EX, 10, 5
MP, NUXY, 3, 0.46

MP, EX, 7, 59
MP, NUXY, 7, 0.49

ESEL, S, TYPE, , 2
EDELE, ALL
ALLSEL

ENORM, 93762
ENORM, 93774
ENORM, 93775
ENORM, 102487
ENORM, 102488
ENORM, 102509
ENORM, 95261
ENORM, 95273
ENORM, 95274
ENORM, 96417
ENORM, 96418
ENORM, 96419
ENORM, 96430
ENORM, 96431
ENORM, 96433
ENORM, 96434
ENORM, 96435
ENORM, 96445
ENORM, 96446
ENORM, 96449
ENORM, 96460
ENORM, 96461
ENORM, 96463
ENORM, 96464
ENORM, 96465
```

```
ENORM, 96475
ENORM, 96476
ENORM, 96477
ENORM, 96478

FINISH

/SOLU
TIME, 1

NSUBST, 100, 3000, 10
OUTRES, ALL, ALL
SOLV
```

Appendix 13: Input File of the 2D Axisymmetrical Poroelastic Knee Model in Section 13.3

```
/PREP7
AA = 22.4/1128
K, 1, (372-372)*AA, -768*AA
K, 2, (1500-372)*AA, -768*AA
K, 3, (1500-372)*AA, -648*AA
K, 4, (372-372)*AA, -648*AA
K, 5, (635-372)*AA, -648*AA
K, 6, (1380-372)*AA, -648*AA
K, 7, (1388-372)*AA, -346*AA
K, 8, (1388-372)*AA, -346*AA
K, 9, (987-372)*AA, -524*AA
K, 10, (1500-372)*AA, -278*AA
K, 11, (1500-372)*AA, -173*AA
K, 12, (1500-372)*AA, -146*AA
K, 13, (372-372)*AA, -146*AA
K, 14, (372-372)*AA, -550*AA
K, 15, (896-372)*AA, -437*AA
K, 16, (1453-372)*AA, -504*AA
K, 17, (1000-372)*AA, -536*AA

BSPLIN, 6, 16, 7
BSPLIN, 4, 9, 8, 10
BSPLIN, 14, 15, 11
BSPLIN, 5, 17, 7
LSTR, 13, 14
LSTR, 12, 13
LSTR, 11, 12
LSTR, 10, 11
LSTR, 4, 14
```

```
LSTR, 5, 6
A, 1, 2, 3, 4

AL, 5, 6, 7, 3
AL, 3, 8, 2, 9
AL, 1, 4, 10

ET, 1, 182
KEYOPT, 1, 3, 1
! BONE MATERIAL
MP, EX, 1, 17600
MP, NUXY, 1, 0.3
! CARTILAGE

ET, 2, 212
KEYOPT, 2, 3, 1
KEYOPT, 2, 12, 1
FPX2 = 3

MP, EX, 2, 0.69,              ! MPA
MP, NUXY, 2, 0.18,

TB, PM, 2, , , PERM
TBDATA, 1, FPX2, FPX2, FPX2
! MENISCUS
FPX3 = 1.26

MP, EX, 3, 0.075,            ! MPA
MP, EY, 3, 0.075
MP, EZ, 3, 100
MP, NUXY, 3, 0.18,
MP, NUXZ, 3, 0.45
MP, NUYZ, 3, 0.45

TB, PM, 3, , , PERM
TBDATA, 1, FPX3, FPX3, FPX3

TYPE, 1
MAT, 1
```

```
ESIZE, 1
AMESH, 2

TYPE, 2
MAT, 2
ESIZE, 1
AMESH, 1, 3, 2

TYPE, 2
MAT, 3
ESIZE, 0.25
AMESH, 4

LSEL, S, LINE, , 13
NSLL, S, 1
CM, N13, NODE
LSEL, S, LINE, , 10
NSLL, S, 1
CM, N10, NODE
LSEL, S, LINE, , 2
NSLL, S, 1
CM, N2, NODE
LSEL, S, LINE, , 4
NSLL, S, 1
CM, N4, NODE
ALLSEL

ET, 7, 169
ET, 8, 171
R, 8
!KEYOPT, 8, 5, 3
KEYOPT, 8, 12, 0
KEYOPT, 8, 1, 8

! GENERATE THE TARGET SURFACE
REAL, 7
TYPE, 7
NSEL, S, , , N13
```

```
CM, _TARGET, NODE
ESLN, S, 0
ESURF
ALLSEL

TYPE, 8
NSEL, S, , , N10
CM, _CONTACT, NODE
ESLN, S, 0
ESURF
ALLSEL

ET, 9, 169
ET, 10, 171
R, 10

KEYOPT, 10, 12, 0
KEYOPT, 10, 1, 8

! GENERATE THE TARGET SURFACE
REAL, 9
TYPE, 9
NSEL, S, , , N13
CM, _TARGET, NODE
ESLN, S, 0
ESURF
ALLSEL

TYPE, 10
NSEL, S, , , N2
CM, _CONTACT, NODE
ESLN, S, 0
ESURF
ALLSEL

ET, 4, 169
ET, 5, 171
R, 5
```

```
KEYOPT, 5, 12, 0
KEYOPT, 5, 1, 8

! GENERATE THE TARGET SURFACE
REAL, 4
TYPE, 4
NSEL, S, , , N2
CM, _TARGET, NODE
ESLN, S, 0
ESURF
ALLSEL
TYPE, 5
NSEL, S, , , N4
CM, _CONTACT, NODE
ESLN, S, 0
ESURF
ALLSEL

LSEL, S, LINE, , 8, 12, 4
LSEL, A, LINE, , 1
NSLL, S, 1
D, ALL, PRES, 0
ALLSEL

LSEL, S, LINE, , 11
NSLL, S, 1
D, ALL, UX, 0
D, ALL, UY, 0
D, ALL, UZ, 0
ALLSEL

LSEL, S, LINE, , 6
NSLL, S, 1
F, ALL, FY, -147/24
ALLSEL
FINISH

/SOLU
```

```
TIME, 1
KBC, 0
NROPT, UNSYM
NSUBST, 100, 1000, 50
OUTRES, ALL, ALL
SOLVE

TIME, 120
KBC, 1
NROPT, UNSYM
NSUBST, 200, 1000, 100
SOLVE
FINI

/POST26
NSOL, 2, NODE(10.462, -12.271, 0), PRES, ,
PLVAR, 2, , , , , , , , , ,
```

Appendix 14: Input File of the Discrete Element Model of Knee Joint in Chapter 14

```
/COM, download dea.cdb from
/COM, www.feabea.net/models/dea.cdb
CDREAD, DB, 'dea', 'cdb'
/PREP7
SECT, 4, SHELL, ,
SECDATA, 0.5
ESEL, S, TYPE, , 2
EMODIF, ALL, SECNUM, 4
ALLSEL

ET, 3, LINK180

! SELECT NODES IN FEMUR WITHIN CONTACT AREA
ESEL, S, MAT, , 4
*GET, NUM0_E, ELEM, 0, COUNT
*GET, E_MIN0, ELEM, 0, NUM, MIN

*DIM, C1, ARRAY, NUM0_E, 4
*DIM, C2, ARRAY, NUM0_E, 4
*DO, I, 1, NUM0_E, 1
  CURR_E = E_MIN0

  *GET, AA, ELEM, CURR_E, AREA
  C1(I, 1) = AA
  C1(I, 2) = CENTRX(CURR_E)
  C1(I, 3) = CENTRY(CURR_E)
  C1(I, 4) = CENTRZ(CURR_E)
  C2(I, 1) = 0.0
  *GET, E_MIN0, ELEM, CURR_E, NXTH
*ENDDO
```

```
/COM,
! BUILD SPRINGS FOR CONTACT AREA
 SHPP, OFF
! THE INITIAL LENGTH IS GIVEN
LTK = 6.5
*DO, I, 1, NUM0_E
  AX = C1(I, 2)
  AY = C1(I, 3)
  AZ = C1(I, 4)
  BX = C1(I, 2)
  BY = C1(I, 3) + 25                                    ! NEED TO ADJUST
  BZ = C1(I, 4)
  ! SELECT NODES IN TIBIA WITHIN CONTACT AREA
  ESEL, S, MAT, , 5
  ESEL, R, CENT, X, AX - 5, AX + 5
  ESEL, R, CENT, Z, AZ - 5, AZ + 5
  *GET, NUM_E, ELEM, 0, COUNT                           ! GET NUMBER OF ELEMENTS
  *GET, E_MIN, ELEM, 0, NUM, MIN                        ! GET MIN ELEMENT NUMBER
  *DO, J, 1, NUM_E
      C_E = E_MIN
       *DO, K, 1, 4
    P0X = CENTRX(C_E)
    P0Y = CENTRY(C_E)
    P0Z = CENTRZ(C_E)
    *IF, K, EQ, 1, THEN
      P1X = NX(NELEM(C_E, 1))
      P1Y = NY(NELEM(C_E, 1))
      P1Z = NZ(NELEM(C_E, 1))
      P2X = NX(NELEM(C_E, 2))
      P2Y = NY(NELEM(C_E, 2))
      P2Z = NZ(NELEM(C_E, 2))
    *ENDIF
    *IF, K, EQ, 2, THEN
      P1X = NX(NELEM(C_E, 2))
      P1Y = NY(NELEM(C_E, 2))
      P1Z = NZ(NELEM(C_E, 2))
      P2X = NX(NELEM(C_E, 3))
      P2Y = NY(NELEM(C_E, 3))
```

```
  P2Z = NZ(NELEM(C_E, 3))
*ENDIF
*IF, K, EQ, 3, THEN
  P1X = NX(NELEM(C_E, 3))
  P1Y = NY(NELEM(C_E, 3))
  P1Z = NZ(NELEM(C_E, 3))
  P2X = NX(NELEM(C_E, 4))
  P2Y = NY(NELEM(C_E, 4))
  P2Z = NZ(NELEM(C_E, 4))
*ENDIF
*IF, K, EQ, 4, THEN
  P1X = NX(NELEM(C_E, 4))
  P1Y = NY(NELEM(C_E, 4))
  P1Z = NZ(NELEM(C_E, 4))
  P2X = NX(NELEM(C_E, 1))
  P2Y = NY(NELEM(C_E, 1))
  P2Z = NZ(NELEM(C_E, 1))
*ENDIF
P0102X = (P1Y - P0Y)*(P2Z - P0Z) - (P1Z - P0Z)*(P2Y - P0Y)
P0102Y = (P1Z - P0Z)*(P2X - P0X) - (P1X - P0X)*(P2Z - P0Z)
P0102Z = (P1X - P0X)*(P2Y - P0Y) - (P1Y - P0Y)*(P2X - P0X)
T1 = P0102X*(AX - P0X) + P0102Y*(AY - P0Y) + P0102Z*(AZ - P0Z)
T2 = -(BX - AX)*P0102X - (BY - AY)*P0102Y - P0102Z*(BZ - AZ)
UV = 100
UV0 = 100
*IF, ABS(T2), GT, 1E-5, THEN
  T = T1/T2
  TX = AX + (BX - AX)*T
  TY = AY + (BY - AY)*T
  TZ = AZ + (BZ - AZ)*T
  P02LX = (P2Y - P0Y)*(AZ - BZ) - (P2Z - P0Z)*(AY - BY)
  P02LY = (P2Z - P0Z)*(AX - BX) - (P2X - P0X)*(AZ - BZ)
  P02LZ = (P2X - P0X)*(AY - BY) - (P2Y - P0Y)*(AX - BX)
  U1 = P02LX*(AX - P0X) + P02LY*(AY - P0Y) + P02LZ*(AZ - P0Z)
  U = U1/T2
  PL01X = (AY - BY)*(P1Z - P0Z) - (AZ - BZ)*(P1Y - P0Y)
  PL01Y = (AZ - BZ)*(P1X - P0X) - (AX - BX)*(P1Z - P0Z)
  PL01Z = (AX - BX)*(P1Y - P0Y) - (AY - BY)*(P1X - P0X)
```

```
        V1 = PL01X*(AX - P0X) + PL01Y*(AY - P0Y) + PL01Z*(AZ - P0Z)
        V = V1/T2
        UV = U + V
    *ENDIF


    *IF, UV, LE, 1.0, THEN
      *IF, U, GT, 0.0, AND, U, LT, 1.0, THEN
        *IF, V, GT, 0.0, AND, V, LT, 1.0, THEN
        UV0 = 0.5
        C2(I, 2) = TX
        C2(I, 3) = TY
        C2(I, 4) = TZ


        N, 1E6 + I, C2(I, 2), C2(I, 3), C2(I, 4)
        N, 2E6 + I, C1(I, 2), C1(I, 3), C1(I, 4)
AB = SQRT((C1(I, 2) - C2(I, 2))**2 + (C1(I, 3) - C2(I, 3))
**2 + (C1(I, 4) - C2(I, 4))**2)
        ISTRAIN = (AB - LTK)/LTK
        SECTYPE, I + 4, LINK,
        SECDATA, C1(I, 1)
        SECCONTROL, , -1
        MAT, 9
        TYPE, 3
        R, 3
        SECNUM, I + 4


        E, 2E6 + I, 1E6 + I
        NSEL, S, NODE, , 1E6 + I
        ESLN, S
        ESEL, R, TYPE, , 3
        INIS, SET, CSYS, -2
        INIS, SET, DTYP, EPEL
        INIS, DEFINE, , , , , ISTRAIN
        ALLSEL
        *ENDIF
       *ENDIF
      *ENDIF
```

```
        *IF, UV0, LE, 1.0, EXIT
     *ENDDO

     *GET, E_MIN, ELEM, C_E, NXTH
     *IF, UV0, LE, 1.0, EXIT
   *ENDDO
*ENDDO

MP, EX, 4, 20E3
MP, NUXY, 4, 0.33

MP, EX, 5, 20E3
MP, NUXY, 5, 0.33

MP, EX, 6, 20E3
MP, NUXY, 6, 0.33

MP, EX, 9, 5
MP, NUXY, 9, 0.33

NSEL, S, NODE, , 1E6, 2E6
D, ALL, UX, 0
D, ALL, UY, -1
D, ALL, UZ, 0
ALLSEL, ALL

MAT, 2
R, 5
REAL, 5
ET, 11, 170
ET, 12, 175
KEYOPT, 12, 9, 1
KEYOPT, 12, 12, 5
KEYOPT, 12, 5, 3
R, 5,
! GENERATE THE TARGET SURFACE
ESEL, S, MAT, , 4
NSLE, S
```

```
TYPE, 11

ESLN, S, 0

ESURF

CMSEL, S, _ELEMCM

! GENERATE THE CONTACT SURFACE

NSEL, S, NODE, , 2E6, 3E6

TYPE, 12

ESLN, S, 0

ESURF

ALLSEL


A = 3E6


ESEL, S, MAT, , 4


! SELECT PART


ET, 7, 185


*GET, NUM_E, ELEM, 0, COUNT        ! GET NUMBER OF ELEMENTS

*GET, E_MIN, ELEM, 0, NUM, MIN     ! GET MIN ELEMENT NUMBER


SHPP, OFF


*DIM, E1, ARRAY, NUM_E, 4


*DO, I, 1, NUM_E, 1                   ! OUTPUT TO ASCII BY LOOPING OVER
                                      ELEMENTS
  CURR_E = E_MIN
  *DO, J, 1, 4
      XX = NX(NELEM(CURR_E, J))
      YY = NY(NELEM(CURR_E, J)) - 1
      ZZ = NZ(NELEM(CURR_E, J))
      N, A + (I - 1)*4 + J, XX, YY, ZZ
      D, A + (I - 1)*4 + J, ALL
  *ENDDO

  TYPE, 7
  MAT, 4
```

```
E, NELEM(CURR_E, 1), NELEM(CURR_E, 2), NELEM(CURR_E, 3),
NELEM(CURR_E, 4), A + (I - 1)*4 + 1, A + (I - 1)*4 + 2, A + (I - 1)
*4 + 3, A + (I - 1)*4 + 4


  *GET, E_MIN, ELEM, CURR_E, NXTH
*ENDDO
ALLSEL
ESEL, S, TYPE, , 2
EDELE, ALL
ALLSEL

FINISH

/SOLU
TIME, 1
NSUBST, 1, 1000, 1
SOLV
FINISH
```

Appendix 15: Input File of the Material Definition of the Cancellous Bone in Chapter 15

/COM, download the whole ankle contact FE model, ankle_contact. cdb, from

/COM, www.feabea.net/models/ankle_contact.cdb

/COM, ankle_contact.cdb should be imported in ANSYS190 and later versions

/COM, THE FOLLOWING INPUT DEFINES THE MATERIAL INTERPOLATION USING RBAS METHOD

/PREP7

! DEFINE ELASTIC MATERIAL INTERPOLATION

TB, ELAS, 5

TBFIELD, XCOR,   191.75

TBFIELD, YCOR,   -22.951

TBFIELD, ZCOR,   135

TBDATA, 1,   111 , 0.30

TBFIELD, XCOR,   185.36

TBFIELD, YCOR,   -33.449

TBFIELD, ZCOR,   135

TBDATA, 1,   134 , 0.30

TBFIELD, XCOR,   185.36

TBFIELD, YCOR,   -33.449

TBFIELD, ZCOR,   145

TBDATA, 1,   120 , 0.30

TBFIELD, XCOR,   177.85

TBFIELD, YCOR,   -44.736

TBFIELD, ZCOR,   135

TBDATA, 1,   83 , 0.30

```
TBFIELD, XCOR,    177.85
TBFIELD, YCOR,    -44.736
TBFIELD, ZCOR,    145
TBDATA,  1,    132 , 0.30

TBFIELD, XCOR,    177.05
TBFIELD, YCOR,    -18.441
TBFIELD, ZCOR,    135
TBDATA,  1,    469 , 0.30

TBFIELD, XCOR,    177.05
TBFIELD, YCOR,    -18.441
TBFIELD, ZCOR,    145
TBDATA,  1,    604 , 0.30

TBFIELD, XCOR,    171.15
TBFIELD, YCOR,    -30.939
TBFIELD, ZCOR,    135
TBDATA,  1,    191 , 0.30

TBFIELD, XCOR,    171.15
TBFIELD, YCOR,    -30.939
TBFIELD, ZCOR,    145
TBDATA,  1,    68 , 0.30

TBFIELD, XCOR,    166.77
TBFIELD, YCOR,    -42.402
TBFIELD, ZCOR,    135
TBDATA,  1,    160 , 0.30

TBFIELD, XCOR,    166.77
TBFIELD, YCOR,    -42.402
TBFIELD, ZCOR,    145
TBDATA,  1,    125 , 0.30

TBFIELD, XCOR,    160.38
TBFIELD, YCOR,    -15.732
TBFIELD, ZCOR,    135
TBDATA,  1,    602 , 0.30

TBFIELD, XCOR,    160.38
```

```
TBFIELD, YCOR,    -15.732
TBFIELD, ZCOR,    145
TBDATA,  1,     357 , 0.30

TBFIELD, XCOR,    155.90
TBFIELD, YCOR,    -29.259
TBFIELD, ZCOR,    135
TBDATA,  1,     192 , 0.30

TBFIELD, XCOR,    155.90
TBFIELD, YCOR,    -29.259
TBFIELD, ZCOR,    145
TBDATA,  1,     393 , 0.30

TBFIELD, XCOR,    151.25
TBFIELD, YCOR,    -40.088
TBFIELD, ZCOR,    135
TBDATA,  1,      90 , 0.30

TBFIELD, XCOR,    151.25
TBFIELD, YCOR,    -40.088
TBFIELD, ZCOR,    145
TBDATA,  1,     158 , 0.30

TBFIELD, XCOR,    149.06
TBFIELD, YCOR,    -15.388
TBFIELD, ZCOR,    135
TBDATA,  1,     403 , 0.30

TBFIELD, XCOR,    149.06
TBFIELD, YCOR,    -15.388
TBFIELD, ZCOR,    145
TBDATA,  1,     659 , 0.30

TBFIELD, XCOR,    145.38
TBFIELD, YCOR,    -25.952
TBFIELD, ZCOR,    135
TBDATA,  1,     407 , 0.30

TBFIELD, XCOR,    145.38
TBFIELD, YCOR,    -25.952
```

```
TBFIELD, ZCOR,     145
TBDATA,  1,      365 , 0.30

TBFIELD, XCOR,     140.79
TBFIELD, YCOR,     -35.642
TBFIELD, ZCOR,     135
TBDATA,  1,      351 , 0.30

TBFIELD, XCOR,     140.79
TBFIELD, YCOR,     -35.642
TBFIELD, ZCOR,     145
TBDATA,  1,      222 , 0.30
! TBIN, ALGO, LMUL
! TBIN, ALGO, NNEI
TBIN, ALGO, RBAS
!***************************************
! DEFINE PLASTIC MATERIAL INTERPOLATION
FF = 0.015
TB, PLASTIC, 5, , , MISO
TBFIELD, XCOR,     191.75
TBFIELD, YCOR,     -22.951
TBFIELD, ZCOR,     135
TBPT, DEFI,  0,    111*FF
TBPT, DEFI,  1E-2, 111*FF

TBFIELD, XCOR,     185.36
TBFIELD, YCOR,     -33.449
TBFIELD, ZCOR,     135
TBPT,   DEFI, 0,   134*FF
TBPT, DEFI, 1E-2,  134*FF

TBFIELD, XCOR,     185.36
TBFIELD, YCOR,     -33.449
TBFIELD, ZCOR,     145
TBPT, DEFI,  0,    120*FF
TBPT, DEFI, 1E-2,  120*FF

TBFIELD, XCOR,     177.85
TBFIELD, YCOR,     -44.736
```

```
TBFIELD, ZCOR,    135
TBPT, DEFI,  0,    83*FF
TBPT, DEFI, 1E-2,   83*FF

TBFIELD, XCOR,    177.85
TBFIELD, YCOR,    -44.736
TBFIELD, ZCOR,    145
TBPT, DEFI, 0  ,   132*FF
TBPT, DEFI, 1E-2,  132*FF

TBFIELD, XCOR,    177.05
TBFIELD, YCOR,    -18.441
TBFIELD, ZCOR,    135
TBPT, DEFI,  0,    469*FF
TBPT, DEFI, 1E-2,  469*FF

TBFIELD, XCOR,    177.05
TBFIELD, YCOR,    -18.441
TBFIELD, ZCOR,    145
TBPT, DEFI,  0,    604*FF
TBPT, DEFI, 1E-2,  604*FF

TBFIELD, XCOR,    171.15
TBFIELD, YCOR,    -30.939
TBFIELD, ZCOR,    135
TBPT, DEFI, 0,    191*FF
TBPT, DEFI, 1E-2,  191*FF

TBFIELD, XCOR,    171.15
TBFIELD, YCOR,    -30.939
TBFIELD, ZCOR,    145
TBPT, DEFI,  0,    68*FF
TBPT, DEFI, 1E-2,   68*FF

TBFIELD, XCOR,    166.77
TBFIELD, YCOR,    -42.402
TBFIELD, ZCOR,    135
TBPT, DEFI,  0,    160*FF
TBPT, DEFI, 1E-2,   160*FF

TBFIELD, XCOR,    166.77
TBFIELD, YCOR,    -42.402
```

```
TBFIELD, ZCOR,    145
TBPT, DEFI,  0,   125*FF
TBPT, DEFI, 1E-2,   125*FF

TBFIELD, XCOR,    160.38
TBFIELD, YCOR,    -15.732
TBFIELD, ZCOR,    135
TBPT, DEFI,  0,   602*FF
TBPT, DEFI, 1E-2,   602*FF

TBFIELD, XCOR,    160.38
TBFIELD, YCOR,    -15.732
TBFIELD, ZCOR,    145
TBPT, DEFI,  0,   357*FF
TBPT, DEFI, 1E-2,   357*FF

TBFIELD, XCOR,    155.90
TBFIELD, YCOR,    -29.259
TBFIELD, ZCOR,    135
TBPT, DEFI,  0,   192*FF
TBPT, DEFI, 1E-2,   192*FF

TBFIELD, XCOR,    155.90
TBFIELD, YCOR,    -29.259
TBFIELD, ZCOR,    145
TBPT, DEFI,  0,   393*FF
TBPT, DEFI, 1E-2,   393*FF

TBFIELD, XCOR,    151.25
TBFIELD, YCOR,    -40.088
TBFIELD, ZCOR,    135
TBPT, DEFI,  0,   90*FF
TBPT, DEFI, 1E-2,   90*FF

TBFIELD, XCOR,    151.25
TBFIELD, YCOR,    -40.088
TBFIELD, ZCOR,    145
TBPT, DEFI,  0,   158*FF
TBPT, DEFI, 1E-2,   158*FF
TBFIELD, XCOR,    149.06
```

```
TBFIELD, YCOR,    -15.388
TBFIELD, ZCOR,    135
TBPT, DEFI,  0,    403*FF
TBPT, DEFI, 1E-2,   403*FF

TBFIELD, XCOR,    149.06
TBFIELD, YCOR,    -15.388
TBFIELD, ZCOR,    145
TBPT, DEFI,  0,    659*FF
TBPT, DEFI, 1E-2,  659*FF

TBFIELD, XCOR,    145.38
TBFIELD, YCOR,    -25.952
TBFIELD, ZCOR,    135
TBPT, DEFI,  0,    407*FF
TBPT, DEFI, 1E-2,   407*FF

TBFIELD, XCOR,     145.38
TBFIELD, YCOR,    -25.952
TBFIELD, ZCOR,    145
TBPT, DEFI, 0,     365*FF
TBPT, DEFI, 1E-2,   365*FF

TBFIELD, XCOR,     140.79
TBFIELD, YCOR,    -35.642
TBFIELD, ZCOR,    135
TBPT, DEFI,  0,    351*FF
TBPT, DEFI, 1E-2,   351*FF

TBFIELD, XCOR,     140.79
TBFIELD, YCOR,    -35.642
TBFIELD, ZCOR,    145
TBPT, DEFI,  0,    222*FF
TBPT, DEFI, 1E-2,   222*FF
! TBIN, ALGO, LMUL
! TBIN, ALGO, NNEI
TBIN, ALGO, RBAS
ALLSEL
```

Appendix 16: Input File of the Stent Implantation Model in Chapter 16

```
/COM, download stent.cdb from
/COM, www.feabea.net/models/stent.cdb
/PREP7
! TURN OFF SHAPE CHECKING BECAUSE CHECKS ALREADY PERFORMED INSIDE
WB MESHER.
! SEE HELP SYSTEM FOR MORE INFORMATION.
SHPP, OFF, , NOWARN
CDREAD, db, 'stent', 'cdb', , '', ''
MP, EX, 1, 53000
MP, NUXY, 1, 0.33
!DEFINE SMA MATERIAL PROPERTIES
TB, SMA, 1, , , SUPE
TBDATA, 1, 345.2, 403, 168, 101, 0.056, 0

C10 = 18.90E-3
C01 = 2.75E-3
C20 = 590.43E-3
C11 = 857.2E-3
NU1 = 0.49
DD = 2*(1 - 2*NU1)/(C10+C01)

TB, HYPER, 2, , 5, MOONEY
TBDATA, 1, C10, C01, C20, C11, , DD

MP, EX, 3, 2.19
MP, NUXY, 3, 0.49
CSYS, 5
WPCSYS, -1
```

```
CYLIND, 1.95, 1.5, 17, 25, 0, 360,
CONE, 1.95, 1.5, 17, 20, 0, 360,
CONE, 1.5, 1.95, 22, 25, 0, 360,
VSBV, 1, 2
VSBV, 4, 3

CYLIND, 2.15, 1.95, 7, 35, 0, 360,
VGLUE, ALL
ET, 10, 187

ESIZE, 0.5, 0,
TYPE, 10
VMESH, ALL

VSEL, S, VOLU, , 1
ESLV, S
EMODIF, ALL, MAT, 3
ALLSEL

VSEL, S, VOLU, , 3
ESLV, S
EMODIF, ALL, MAT, 2
ALLSEL

ASEL, S, AREA, , 5, 7
ASEL, A, AREA, , 15, 16
ASEL, A, AREA, , 1
NSLA, S, 1
NPLOT
CM, TARGET, NODE
ALLSEL

MAT, 4
R, 3
REAL, 3
ET, 3, 170
ET, 4, 175
```

```
KEYOPT, 4, 9, 1
KEYOPT, 4, 10, 0
R, 3,
RMORE,
RMORE, , 0
RMORE, 0
! GENERATE THE TARGET SURFACE
NSEL, S, , , TARGET
CM, _TARGET, NODE
TYPE, 3
ESLN, S, 0
ESURF
CMSEL, S, _ELEMCM
! GENERATE THE CONTACT SURFACE
NSEL, S, , , CONTACT
CM, _CONTACT, NODE
TYPE, 4
ESLN, S, 0
ESURF


!USE FORCE-DISTRIBUTED BOUNDARY CONSTRAINTS ON 2 SIDES OF ARTERY
WALL
! TO ALLOW FOR RADIAL EXPANSION OF TISSUE WITHOUT RIGID BODY MOTION
MAT, 4
R, 15
REAL, 15
ET, 17, 170
ET, 18, 175
KEYOPT, 18, 12, 5
KEYOPT, 18, 4, 1
KEYOPT, 18, 2, 2
KEYOPT, 17, 2, 1
KEYOPT, 17, 4, 111111
TYPE, 17
! CREATE A PILOT NODE
N, 100001, 0, 0, 35
TSHAP, PILO
```

```
E, 100001
! GENERATE THE CONTACT SURFACE
LSEL, S, , , 35, 38
CM, _CONTACT, LINE
TYPE, 18
NSLL, S, 1
ESLN, S, 0
ESURF

MAT, 4
R, 14
REAL, 14
ET, 15, 170
ET, 16, 175
KEYOPT, 16, 12, 5
KEYOPT, 16, 4, 1
KEYOPT, 16, 2, 2
KEYOPT, 15, 2, 1
KEYOPT, 15, 4, 111111
TYPE, 15
! CREATE A PILOT NODE
N, 100002, 0, 0, 7
TSHAP, PILO
E, 100002
! GENERATE THE CONTACT SURFACE
LSEL, S, , , 19, 22
CM, _CONTACT, LINE
TYPE, 16
NSLL, S, 1
ESLN, S, 0
ESURF

MAT, 4
R, 13
REAL, 13
ET, 13, 170
ET, 14, 175
```

```
KEYOPT, 14, 12, 5
KEYOPT, 14, 4, 1
KEYOPT, 14, 2, 2
KEYOPT, 13, 2, 1
KEYOPT, 13, 4, 111111
TYPE, 13
! CREATE A PILOT NODE
N, 100003, 0, 0, 18.268
TSHAP, PILO
E, 100003
! GENERATE THE CONTACT SURFACE

TYPE, 14
ESEL, S, MAT, , 1
NSLE, S
NSEL, R, LOC, Z, 18.268
ESLN, S, 0
ESURF

MAT, 4
R, 12
REAL, 12
ET, 11, 170
ET, 12, 175
KEYOPT, 12, 12, 5
KEYOPT, 12, 4, 1
KEYOPT, 12, 2, 2
KEYOPT, 11, 2, 1
KEYOPT, 11, 4, 111111
TYPE, 11
! CREATE A PILOT NODE
N, 100004, 0, 0, 23.5
TSHAP, PILO
E, 100004
! GENERATE THE CONTACT SURFACE

TYPE, 12
ESEL, S, MAT, , 1
```

```
NSLE, S
NSEL, R, LOC, Z, 23.5
ESLN, S, 0
ESURF
ALLSEL

D, 100001, ALL
D, 100002, ALL
D, 100003, ALL
D, 100004, ALL
ALLSEL
ASEL, S, AREA, , 5, 7
ASEL, A, AREA, , 15, 16
ASEL, A, AREA, , 1
NSLA, S, 1
SF, ALL, PRES, 0.15
ALLSEL

ASEL, S, AREA, , 13, 14
ASEL, A, AREA, , 17, 18
NSLA, S, 1
SF, ALL, PRES, 0.04
ALLSEL
FINISH

/SOLU
ANTYPE, 0
NLGEOM, ON
!APPLY LOAD STEP 1: EXPAND VESSEL PAST THE RADIUS OF THE STENT BY
APPLYING
! BIG PRESSURE ON VESSEL

NSUBST, 20, 1000
CNCHECK, AUTO
ESEL, S, TYPE, , 4
CM, CONTACT2, ELEM
```

```
EKILL, CONTACT2 !KILL CONTACT ELEMENTS IN STENT-PLAQUE CONTACT
PAIR
ALLSEL

SOLVE


!APPLY LOAD STEP 2: ACTIVATE CONTACT BETWEEN STENT AND PLAQUE
!ACTIVATE STENT ELEMENTS
EALIVE, CONTACT2
ALLS

NSUBST, 2, 2
SOLVE


!APPLY LOAD STEP 3: REDUCE THE PRESSURE ON THE WALL TO 13.3 KPA
! TO ALLOW THE STENT CONTACTING THE VESSEL AND PLAQUE
NSUBST, 300, 3000, 30
ASEL, S, AREA, , 5, 7
ASEL, A, AREA, , 15, 16
ASEL, A, AREA, , 1
NSLA, S, 1
SF, ALL, PRES, 0.0133
ALLS
ASEL, S, AREA, , 13, 14
ASEL, A, AREA, , 17, 18
NSLA, S, 1
SF, ALL, PRES, 0.0133
ALLSEL

SOLVE
SAVE
FINISH
```

Appendix 17: Input File of the Wear Model of Hip Replacement in Chapter 17

```
/COM, download wear_model.dat from
/COM, www.feabea.net/models/wear_model.dat
/COM, this model can be repeated in ANSYS190 and later versions
/PREP7
! TURN OFF SHAPE CHECKING
SHPP, OFF, , NOWARN
/NOLIST
/INP, wear_model, dat
ALLSEL
/COM,
LOCAL,  12,  0,  -25.9412130805917,  -3.2275027035005,  4.5,
-60.0139777423635, 0., 180.
CSYS, 0


MP, EX, 1, 58000,
MP, NUXY, 1, 0.3,


MP, EX, 2, 58000,
MP, NUXY, 2, 0.3,


MP, EX, 3, 1200,
MP, NUXY, 3, 0.4,


MP, EX, 4, 58000,
MP, NUXY, 4, 0.3,


KUW = 30E-7*1E6*3.1416*25 / 1000
```

```
TB, WEAR, 5, , , ARCD      ! MAT5 FOR WEAR OF CONTACT ELEMENT ON LINER
TBFIELD, TIME, 0
TBDATA, 1, 0, 1, 1, 0, 0    ! NO WEAR FOR LOAD STEP#1;
TBFIELD, TIME, 1
TBDATA, 1, 0, 1, 1, 0, 0
TBFIELD, TIME, 1.01         ! START WEAR IN LOAD STEP #2
TBDATA, 1, KUW, 1, 1, 0, 0
TBFIELD, TIME, 2
TBDATA, 1, KUW, 1, 1, 0, 0

ESEL, S, TYPE, , 5
EMODIF, ALL, MAT, 5
ALLSEL

! DEFINE COMPONENT FOR NLAD
ALLSEL, ALL, ALL
ESEL, S, TYPE, , 5, 5
CM, CONWEAREL, ELEM
ALLSEL, ALL, ALL

NSEL, S, LOC, Y, -200, -134
D, ALL, ALL
ALLSEL

ET, 12, 170

ET, 14, 174                 ! REMOTE FORCE CONTACT
KEYO, 14, 4, 2
KEYO, 14, 2, 2        ! MPC
KEYO, 14, 12, 5       ! BONDED CONTACT

CMSEL, S, LOADINGSURFACE, NODE
TYPE, 14
REAL, 14
ESURF
TYPE, 12
N, 20000, -47.737, 37.460+10, -6.9274
```

```
TSHAP, PILOT
E, 20000

CSYS, 12
NROTAT, 20000
D, 20000, UY
D, 20000, UZ
F, 20000, FX, -2000
ALLSEL

CSYS, 0
FINISH


/COM, ==================================
/COM, SOLUTION
/COM, ==================================
/SOLU

! LOAD STEP 1: APPLY NODAL FORCES, NO WEAR
NLGEOM, ON
ALLSEL, ALL
OUTRES, ALL, ALL
! DEFINE NON-LINEAR ADAPTIVITY CRITERION
NLAD, CONWEAREL, ADD, CONTACT, WEAR, 0.50    ! MORPH AFTER 50 % IS
                                             LOST IN WEAR
NLAD, CONWEAREL, ON, ALL, ALL, 1, , 2
NLAD, CONWEAREL, LIST, ALL, ALL
TIME, 1
DELTIM, 0.1, 1E-4, 1
SOLVE

! LOAD STEP 2: START THE WEAR
TIME, 2
DELTIM, 0.01, 1E-6, 0.02
NLHIST, PAIR, WV, CONT, WEAR, 5
SOLVE
FINISH
```

```
/POST1
ESEL, S, TYPE, , 5
NSLE, S
ESLN, S
ESEL, R, MAT, , 5
ETABLE, WEAR, NMISC, 176
PLETAB, WEAR
```

Appendix 18: Input File of the Mini Dental Implant Crack-Growth Model in Chapter 18

```
/COM, download dental_model.dat from
/COM, www.feabea.net/models/dental_model.dat
/COM, this model can be repeated in ANSYS190 and later versions
/PREP7
! TURN OFF SHAPE CHECKING BECAUSE CHECKS ALREADY PERFORMED INSIDE
WB MESHER.
! SEE HELP SYSTEM FOR MORE INFORMATION.
SHPP, OFF, , NOWARN
/NOLIST
/INP, dental_model, dat
! PARIS' LAW CONSTANTS
C = 2.29E-10
M = 2

! FATIGUE CRACK GROWTH LAW SPECIFICATION
TB, CGCR, 2, , , PARIS
TBDATA, 1, C, M

MP, EX, 1, 200000
MP, NUXY, 1, 0.33

/COM, ********* START CREATING ASSIST NODES FOR CRACK
CALCULATION ***********
N, 115121, -6.95106442935247E-025, 0.558242785259289,
-2.00000003678724
/COM, *********** DONE CREATING ASSIST NODES FOR CRACK
CALCULATION ***********
```

```
/WB, LOAD, START                     ! STARTING TO SEND LOADS
NSEL, S, LOC, Z, 7, 12
D, ALL, ALL
ALLSEL
NSEL, S, NODE, , NODE(0.4115, -0.25332, -5.3625)
NSEL, A, NODE, , NODE(-0.23072, -0.24340, -5.4489)
NSEL, A, NODE, , NODE(-0.062519, 0.26226, -5.4602)

F, ALL, FY, -33.3
ALLSEL
LOCAL, 13, 0, -6.95106442935247E-025, 0.558242785259289,
-2.00000003678724, 0., -90., 90.

CSYS, 0

FINISH
/COM, *********** START CINT COMMANDS FOR ALL CRACKS ***********
/SOLU
ANTYPE, 0
*SET, _IASSISTNODE, 115121
*SET, _SIFS, 1
CINT, NEW, 1              ! DEFINE CRACK ID FOR SEMI-ELLIPTICAL CRACK
CINT, TYPE, SIFS         ! OUTPUT QUANTITY FOR SEMI-ELLIPTICAL CRACK
CINT, CTNC, NS_SECRACK_FRONT, _IASSISTNODE ! DEFINE CRACK TIP
NODE COMPONENT
CINT, NCON, 6                         ! DEFINE NUMBER OF CONTOURS
CINT, NORMAL, 13, 2                   ! DEFINE CRACK PLANE NORMAL
! DEFINE CRACK TOP AND BOTTOM SURFACES

CINT, SURF, NS_SECRACK_TOPFACE, NS_SECRACK_BOTTOMFACE

CGROW, NEW, 1
CGROW, CID, 1
CGROW, METHO, SMART, REME

CGROW, FCG, METH, LC                  ! CYCLE BY CYCLE METHOD
CGROW, FCG, DAMX, 0.1                 ! MAXIMUM CRACK GROWTH INCREMENT
CGROW, FCG, SRAT, 0                   ! STRESS-RATIO
CGROW, FCOPTION, MTAB, 2
```

```
NSUBST, 5, 5, 5
OUTRES, ALL, ALL

SOLV
FINISH
/POST1

 *GET, NSTEP, ACTIVE, 0, SET, NSET
 CRKID = 1

 MAXNUMND = 0
 SET, FIRST
 *DO, ISTEP, 1, NSTEP
 ! GET NUMBER OF CRACK TIPS
   *GET, FVAL, CINT, 1, NNOD
    *IF, FVAL, GT, MAXNUMND, THEN
      MAXNUMND = FVAL
    *ENDIF
 SET, NEXT
 *ENDDO

 *DIM, NUMND, ARRAY, NSTEP, 1
 *DIM, NODENUM, ARRAY, MAXNUMND, NSTEP
 *DIM, DTN, ARRAY, 2, NSTEP
 *DIM, DTA, ARRAY, 2, NSTEP
 *DIM, DTK, ARRAY, 2, NSTEP
 *DIM, DTR, ARRAY, 2, NSTEP
 *DIM, ND_X, ARRAY, 2, NSTEP
 *DIM, ND_Y, ARRAY, 2, NSTEP
 *DIM, ND_Z, ARRAY, 2, NSTEP
 *DIM, ND_A, ARRAY, 2, NSTEP
 ! RST0 STORES THE RESULTS OF THE FIRST NODE
 ! RST1 STORES THE RESULTS OF THE LAST NODE
 *DIM, RST0_DTN, ARRAY, NSTEP, 2
 *DIM, RST0_DTA, ARRAY, NSTEP, 2
 *DIM, RST0_DTK, ARRAY, NSTEP, 2
 *DIM, RST1_DTN, ARRAY, NSTEP, 2
```

```
*DIM, RST1_DTA, ARRAY, NSTEP, 2
*DIM, RST1_DTK, ARRAY, NSTEP, 2

SET, FIRST
! GET NUMBER OF CRACK TIPS
*GET, FVAL, CINT, 1, NNOD
ISTEP = 1
NUMND(ISTEP, 1) = FVAL
! GET TIP NODE NUMBERS
*DO, INODE, 1, 2
  *IF, INODE, EQ, 1, THEN
    *GET, NDNUM, CINT, CRKID, NODE, INODE
  *ELSE
    *GET, NDNUM, CINT, CRKID, NODE, NUMND(ISTEP, 1)
  *ENDIF

NODENUM(INODE, ISTEP) = NDNUM

*GET, FVAL, CINT, CRKID, CTIP, NDNUM, CONTOUR, 1, DTYPE, DLTN
DTN(INODE, ISTEP) = FVAL
*GET, FVAL, CINT, CRKID, CTIP, NDNUM, CONTOUR, 1, DTYPE, DLTA
DTA(INODE, ISTEP) = FVAL
*GET, FVAL, CINT, CRKID, CTIP, NDNUM, CONTOUR, 1, DTYPE, DLTK
DTK(INODE, ISTEP) = FVAL
*GET, FVAL, CINT, CRKID, CTIP, NDNUM, CONTOUR, 1, DTYPE, R
DTR(INODE, ISTEP) = FVAL

*GET, FVAL, CINT, CRKID, CTIP, NDNUM, CONTOUR, 1, DTYPE, CRDX
ND_X(INODE, ISTEP) = FVAL
*GET, FVAL, CINT, CRKID, CTIP, NDNUM, CONTOUR, 1, DTYPE, CRDY
ND_Y(INODE, ISTEP) = FVAL
*GET, FVAL, CINT, CRKID, CTIP, NDNUM, CONTOUR, 1, DTYPE, CRDZ
ND_Z(INODE, ISTEP) = FVAL
*GET, FVAL, CINT, CRKID, CTIP, NDNUM, CONTOUR, 1, DTYPE, APOS
ND_A(INODE, ISTEP) = FVAL

  *IF, INODE, EQ, 1, THEN
   RST0_DTN(ISTEP, 1) = DTN(1, ISTEP)
```

```
      RST0_DTA(ISTEP, 1) = DTA(1, ISTEP)
      RST0_DTK(ISTEP, 1) = DTK(1, ISTEP)

  *ELSE
      RST1_DTN(ISTEP, 1) = DTN(INODE, ISTEP)
      RST1_DTA(ISTEP, 1) = DTA(INODE, ISTEP)
      RST1_DTK(ISTEP, 1) = DTK(INODE, ISTEP)

  *ENDIF
 *ENDDO

SET, NEXT

*DO, ISTEP, 2, NSTEP

! GET NUMBER OF CRACK TIPS
  *GET, FVAL, CINT, 1, NNOD
  NUMND(ISTEP, 1) = FVAL
! GET TIP NODE NUMBERS
  *DO, INODE, 1, 2
    *IF, INODE, EQ, 1, THEN
     *GET, NDNUM, CINT, CRKID, NODE, INODE
    *ELSE
     *GET, NDNUM, CINT, CRKID, NODE, NUMND(ISTEP, 1)
    *ENDIF

  *GET, FVAL, CINT, CRKID, CTIP, NDNUM, CONTOUR, 1, DTYPE, DLTN
  DTN(INODE, ISTEP) = DTN(INODE, ISTEP - 1) + FVAL
  *GET, FVAL, CINT, CRKID, CTIP, NDNUM, CONTOUR, 1, DTYPE, DLTA
  DTA(INODE, ISTEP) = DTA(INODE, ISTEP - 1) + FVAL
  *GET, FVAL, CINT, CRKID, CTIP, NDNUM, CONTOUR, 1, DTYPE, DLTK
  DTK(INODE, ISTEP) = FVAL
  *GET, FVAL, CINT, CRKID, CTIP, NDNUM, CONTOUR, 1, DTYPE, R
  DTR(INODE, ISTEP) = FVAL

  *GET, FVAL, CINT, CRKID, CTIP, NDNUM, CONTOUR, 1, DTYPE, CRDX
  ND_X(INODE, ISTEP) = FVAL
  *GET, FVAL, CINT, CRKID, CTIP, NDNUM, CONTOUR, 1, DTYPE, CRDY
  ND_Y(INODE, ISTEP) = FVAL
```

```
*GET, FVAL, CINT, CRKID, CTIP, NDNUM, CONTOUR, 1, DTYPE, CRDZ
ND_Z(INODE, ISTEP) = FVAL
*GET, FVAL, CINT, CRKID, CTIP, NDNUM, CONTOUR, 1, DTYPE, APOS
ND_A(INODE, ISTEP) = FVAL

*IF, INODE, EQ, 1, THEN
  RST0_DTN(ISTEP, 1) = DTN(1, ISTEP)
  RST0_DTA(ISTEP, 1) = DTA(1, ISTEP)
  RST0_DTK(ISTEP, 1) = DTK(1, ISTEP)

*ELSE
  RST1_DTN(ISTEP, 1) = DTN(INODE, ISTEP)
  RST1_DTK(ISTEP, 1) = DTK(INODE, ISTEP)
  RST1_DTA(ISTEP, 1) = DTA(INODE, ISTEP)

*ENDIF
*ENDDO

SET, NEXT

*ENDDO

*DIM, LABEL, ARRAY, NSTEP, 1
*DIM, DNTAB, TABLE, NSTEP, 1
*DIM, DATAB, TABLE, NSTEP, 1
*DIM, DKTAB, TABLE, NSTEP, 1

*DO, I, 1, NSTEP
  LABEL(I) = I
  *VFILL, DNTAB(I, 1), DATA, DTN(1, I)
  *VFILL, DATAB(I, 1), DATA, DTA(1, I)
  *VFILL, DKTAB(I, 1), DATA, DTK(1, I)
*ENDDO

/OUT

/COM, RESULTS OF THE STARTING POINT AT THE CRACK FRONT
/COM, | NSTEP | DTK | DTA | DTN |
*VWRITE, LABEL(1), DATAB(1, 1), DKTAB(1, 1), DNTAB(1, 1)
(3X, F14.5, ' ', F14.8, ' ', F14.8, ' ', F14.5)
```

Index

Printed and bound by CPI Group (UK) Ltd, Croydon, CR0 4YY

17/10/2024

01775660-0011